T0406300

Environmental Science and Engineering

Environmental Science

More information about this series at http://www.springer.com/series/3234

Walter Leal Filho · Marina Kovaleva

Food Waste and Sustainable Food Waste Management in the Baltic Sea Region

Walter Leal Filho
Marina Kovaleva
Life Sciences
Hamburg University of Applied Sciences
Hamburg
Germany

ISSN 1431-6250
ISBN 978-3-319-10905-3 ISBN 978-3-319-10906-0 (eBook)
DOI 10.1007/978-3-319-10906-0

Library of Congress Control Number: 2014949358

Springer Cham Heidelberg New York Dordrecht London

Printed on acid-free paper

Springer is part of Springer Science+Business Media (www.springer.com)

Preface

Today, the amount of food thrown away worldwide, reaches around 1.3 billion tonnes per year. This book presents the findings of an extensive piece of research on the state of the problem of food waste in Belarus, Estonia, Germany, Latvia, Lithuania, Poland and Sweden. The results show that the scale of the problem with regard to food waste varies between each country and is limited by an insufficient number of studies in the area. In all countries except Germany and Sweden, the problem is most prevalent in the area of food waste generated by the manufacturing sector, mostly stemming from unused or inefficient use of by-products. In Germany and Sweden, the main problem is food thrown away by households that is still suitable for human consumption. The values reach 47–65 % and 35 %, respectively. The method to reduce or prevent food waste most often applied across the seven countries is the donation of food. In addition, Germany has initiated a large number of engagement campaigns and activities aimed at reduction of food waste, whereas, Sweden has launched projects only focused on single organisations or institutions. The other reduction and prevention methods are similar to those used for biodegradable waste in the countries included in this study. The results gathered in this study show some potential measures/methods and areas, which may be considered in future work in order to reduce the amount of food waste generated in each of the countries included in the study.

The authors would like to thank, the Estonian Food Bank; Federation of Polish Food Banks; Center for Environmental Solutions in Belarus; Sustainable Business Hub in Malmö, Sweden; Latvian Food Bank 'Paēdušai Latvijai'; Lithuanian Food Bank 'Maisto bankas'; Kieler Tafel in Germany; European Federation of Food Banks and 'Hanzas Maiznīcas' company in Latvia, who have willingly shared their time to provide data and assist with this study.

Contents

Abbreviations

ABPR	Animal By-Product Regulations
AD	Anaerobic digestion
BAT	Best Available Technology
BMELV	German Federal Ministry of Food Agriculture and Consumer Protection
BMW	Biodegradable Municipal Waste
BOGOF	Buy one get one free
CBI	Centre for the Promotion of Imports from developing countries
CEWEP	Confederation of European Waste-to-Energy Plants
CHP	Combined Heat and Power
Defra	Department for Environment, Food and Rural Affairs
EAUC	Environmental Association for Universities and Colleges
EC	European Commission
ECN	European Compost Network
EEA	European Environment Agency
EU	European Union
EWWR	European Week for Waste Reduction
FAO	Food and Agriculture Organization
FEBA	European Federation of Food Banks
FFV	Fresh Fruits and Vegetables
FSC	Food Supply Chain
GDP	Gross Domestic Product
GHG	Greenhouse Gases
IEEP	Institute for European Environmental Policy
IES	Institute for Environment and Sustainability
ISO	International Standard Organisation
ISWM	Integrated Solid Waste Management
IVC	In-vessel composting
JRC	Joint Research Centre, Institute for Environment and Sustainability
MBT	Mechanical-Biological Treatment
MRL	Maximum Residue Level

MS	Member States
MSW	Municipal Solid Waste
NSW	New South Wales
RDF	Refuse-Derived Fuel
RFID	Radio Frequency Identification technology
SHR	Swedish Hotel and Restaurant Association
TPR	Temporary Price Reduction
US EPA	United States Environmental Protection Agency
UK	United Kingdom
UN	United Nations
UNECE	United Nations Economic Commission for Europe
US	United States
Vito	Vision on Technology
WFD	Waste Framework Directive
WRAP	Waste and Resources Action Programme

Chapter 1
Introduction

The rapidly changing world also has a great impact on food production and consumption patterns. Attitudes of society towards food has shifted over the years due to rising income per capita, demographic shifts, changing lifestyles, and moral and social values. Technological innovations and competition in the international food market have driven changes in the variety and availability of food products (BIO Intelligence Service et al. 2011). Nevertheless, the issue of food accessibility and affordability still remains as topical today as it did decades ago. Today, globally, 9 million people die of hunger each year, and 800 million are undernourished (BIO Intelligence Service et al. 2011).

At the same time, according to the FAO estimations, approximately 30 % of all food produced for human consumption is lost or wasted throughout the global food supply system (from initial agricultural production to final household consumption). Food waste amounts to approximately 1.3 billion tonnes per year (Gustavsson et al. 2011). Breaking it down into different food categories, globally, roughly 30 % of cereals, 40–50 % of root crops, fruits and vegetables, 20 % of oilseeds, meat and dairy, and 30 % of fish are discarded annually (FAO 2012b). Worldwide, retailers throw away 1.6 million tonnes of food per year (Institution of Mechanical Engineers 2013).

In medium- and high-income countries food is to a significant extent rejected at the consumption stage due to wasteful behaviour by consumers, as a result of an excessive amount of purchased food. In low-income countries food is mostly lost or wasted during the early and middle stages of the food supply chain (e.g. harvesting, transportation) and much less at the consumer level. However, overall, on a per capita basis, much more food is thrown away in the industrialized world than in developing countries (Gustavsson et al. 2011).

Such wasteful behaviour jeopardises not only the current, but also the future state of food security in the world. This becomes evident in the light of the projected 60 % increase in the global demand for food by 2050, effects of climate change, natural resource constraints (e.g. water scarcity), losses in yield and land area as a result of environmental degradation, and competing demands, especially, for the production of biofuels (Nellemann et al. 2009; FAO et al. 2012). Today, 60 % of the world's major ecosystems have already been degraded or are used unsustainably

W. Leal Filho and M. Kovaleva, *Food Waste and Sustainable Food Waste Management in the Baltic Sea Region*, Environmental Science and Engineering, DOI 10.1007/978-3-319-10906-0_1

(European Commission 2011). The demand for food will also be driven by global population growth: a larger number of wealthier people and required additional resources to produce products for their more varied, high-quality diet (Foresight 2011).

In addition, food which is grown and produced but uneaten has significant environmental and economic costs (FAO 2013). It leads to waste of resources used in production, such as land, water, energy, fertilizers, as well as to unnecessary CO_2 emissions, and has a direct and negative impact on the income of both farmers and consumers (Gustavsson et al. 2011; Institution of Mechanical Engineers 2013). At the European level alone, at least 170 million tonnes of CO_2eq. (approximately 3 % of total EU-27 emissions in 2008) are emitted annually, along all steps of the life cycle of disposed of food, namely agricultural steps, food processing, transportation, storage, consumption steps and end-of-life impacts (BIO Intelligence Service et al. 2011). Moreover, conservative estimates of water loss caused by discarded food indicate that about half of the water withdrawn for irrigation is lost (World Economic Forum 2009).

The direct economic cost of lost or wasted agricultural products (excluding fish and seafood), based on producer prices only, is approximately EUR 548 billion (USD 750 billion), which is equivalent to the GDP of Switzerland (FAO 2013). US businesses and consumers lose about EUR 145 billion (USD 198 billion) per year because of discarded food (Venkat 2011). In the UK thrown away food which is suitable for human consumption costs EUR 12.4 billion (£10.2 billion) per year (WRAP 2008).

The exact causes of rejected food are significantly dependent on the conditions and local situation experienced by a country (Gustavsson et al. 2011). For instance, in low-income countries, these causes are mainly connected to financial, managerial and technical limitations in harvesting techniques, storage and cooling facilities in difficult climatic conditions, infrastructure, packaging and marketing systems (Gustavsson et al. 2011).

Whereas in medium/high-income countries the causes relate to consumer behaviour (e.g. insufficient purchase planning, confusion of date labels, lack/insufficient knowledge/information), quality standards (e.g. not perfect shape, size, colour or time to ripeness of a food item), legislation, a lack of coordination between different actors in the supply chain that leads to oversupply and overproduction, technical malfunctions and challenges to forecast consumer demand.

Unfortunately, the retail model views food disposal as a necessary part of the business (Gunders 2012).

In the area of food service, the causes of food waste are large portion sizes and undesired accompaniments, inflexibility of chain-store management and pressure to maintain enough food supply to offer extensive menu choices at all times (Gunders 2012).

The available statistics regarding amounts of discarded food in a single county or region is 'impressive'. USA, Canada, Australia, and New Zealand collectively dispose of 38 % of grain products, 50 % of seafood, 52 % of fruits and vegetables, 22 % of meat, and 20 % of milk (Gunders 2012).

According to FAO, lost or wasted food per capita in Europe and North-America amounts to 280–300 kg per year (Gustavsson et al. 2011). The European studies bring a value of 179 kg per capita that in total comprise 89 million tonnes (BIO Intelligence Service et al. 2011).

In developing countries, 35–50 % of lost or wasted food is caused by inefficiencies in the entire value chain of food products (mainly: harvesting; storage; transportation and processing stages; World Economic Forum 2009).

In Asia, these amount to 10–37 % for cereals and oilseed, and to approximately 50 % for some perishable staples (World Economic Forum 2009).

In the United States approximately 7 % of planted fields are typically not harvested each year (Gunders 2012).

In the EU, the manufacturing sector generates 39 % of the total of food related waste, or approximately 35 million tonnes (BIO Intelligence Service et al. 2011) which is almost the same amount as in the USA—36.3 million tonnes (U.S. EPA 2013).

In the industrialised countries, the amount of food that is discarded by retail, food service and household sectors raise the biggest concern. In 2008, in-store food loss or waste in the United States was estimated to be 19.5 million tonnes: equivalent to 10 % of the total food supply at the retail level. Approximately 4–10 % of food purchased by restaurants becomes kitchen loss, both edible and inedible, before reaching the consumer (Gunders 2012). In the EU-27, the wholesale/retail sector generates close to 8 kg of food loss or waste per capita, representing around 4.4 million tonnes per year. The food service sector generates an average of 25 kg per capita, 12.3 million tonnes overall (BIO Intelligence Service et al. 2011).

At consumer level, the industrialised countries discard about 222 million tonnes, which is almost as high as the total net food production in sub-Saharan Africa (230 million tonnes) (Gustavsson et al. 2011).

A consumer in Europe and North America discards on average between 95 and 115 kg per year, while in sub-Saharan Africa and South/Southeast Asia a consumer will only discard of 6–11 kg per year on average (Gustavsson et al. 2011).

In the United States 40 % of food goes uneaten. Today, the average American consumer wastes up 50 % more food than American consumers in the 1970s. American families throw out approximately 25 % of purchased food (Gunders 2012). The same value is true for consumers in the UK.

In the EU, households produce the largest fraction of food related waste overall, at about 42 % of the total or about 38 million tonnes (BIO Intelligence Service et al. 2011). A detailed country-level study conducted in the UK showed that 61 % or 4.1 million tonnes of food are discarded because it had not been managed well. 46 % of the wasted food is in a fresh, raw or minimally processed state, 27 % having been cooked or prepared in some way and 20 % ready to consume when purchased.

45 thousand tonnes of rice, 33 thousand tonnes of pasta and 105 thousand tonnes of potato are thrown away each year, suggesting people prepare too much. Over one quarter (nearly 1.2 million tonnes per year) of food is discarded in its packaging, either opened or unopened. Annually, 2.9 billion whole and untouched fruit items,

1.9 billion whole vegetables and 1.2 million bakery items are thrown away (WRAP 2008).

The experts claim that only in the EU, the total amount of discarded food expected to rise by 40 % by 2020 (European Parliament Resolution (2011/2175 (INI)) 2012). Addressing the problem, the EU and UN have signed the 'Joint Declaration Against Food Waste', where they commit to reduce the amount of wasted food by 50 % by 2025 (Weber et al. 2011). In addition, the European Parliament designated 2014 as the 'European year against food waste' (Gunders 2012). It is noted that halving the current global amount of discarded food could reduce the food required by 2050 by an amount approximately equal to 25 % of today's production (Foresight 2011).

However, despite a number of studies undertaken which make the case for tackling the problem, actual precise data on food loss and waste generation, avoidance and management and its ultimate fate is scarce, sparse, fragmented, disaggregated and difficult to verify, both on the global and national levels. This indicates the need to conduct additional research in the area.

The goal of this book is to investigate and make a thorough review of the state of the problem of food waste in the following countries in the Baltic Sea Region: Belarus; Estonia; Latvia; Lithuania; Poland; Germany and Sweden. It includes an analysis of the following aspects: the amount of food waste generated; its composition; stages in the food supply chain where the biggest quantities are accumulated; causes and applied treatment methods. Finally, the thesis will provide a suggestion of possible measures necessary to be taken in order to reduce the amount of food waste generated, based on the obtained results.

The seven countries mentioned above represent differing economies with differing consumer purchasing power. Therefore, the data on the amount of food waste generated in each country gives a foundation to support or refute a *hypothesis* that there is a strong negative dependence between an amount of household food waste and a share of food expenditures, i.e., consumers who spend a smaller amount on food generated a larger amount of food waste.

1.1 Scope

The complexity and, to a lesser extent, the ambiguity of the term 'food waste' makes it difficult to bring the results of many related studies to one common basis, and reduces the possibility for comparison. Confusion around the definition of the term 'food waste', the lack of a harmonized version of the term and the need to establish such a term are also noted in the resolution of the European Parliament (2011/2175(INI)) (European Parliament Resolution (2011/2175(INI)) 2012). In order to define the scope of this work, firstly, it is necessary to build a topology of what today is referred to 'thrown food'. An analysis of available studies and reports has shown that researchers define wasted food by using a number of crossover,

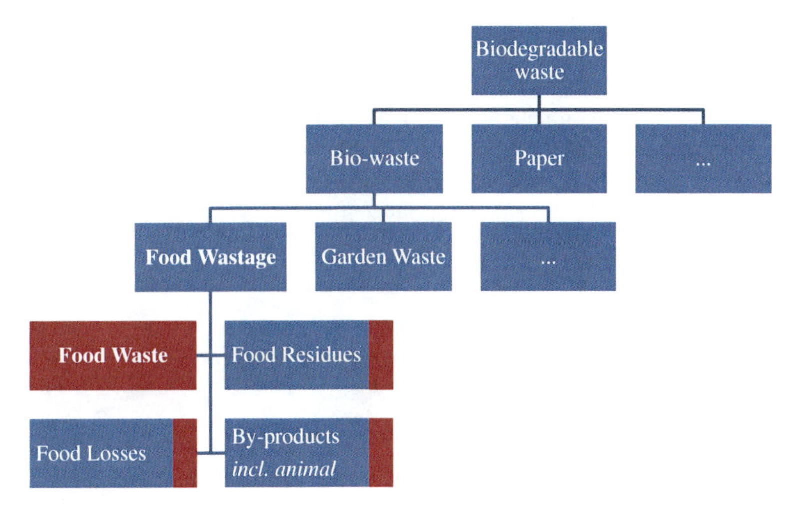

Fig. 1.1 Biodegradable waste hierarchy

interchangeable or, mutually exclusive terms such as 'food loss/es', 'food waste', 'food wastage', 'food residues', 'bio-waste', 'biodegradable waste' etc.

The topology below (Fig. 1.1) is built based on a foundation in the literature terms and definitions regarding discarded food.

At the top of the hierarchy is biodegradable waste, of which one of the constituents is bio-waste. Bio-waste could be divided into a number of sub-types. One of these is 'food wastage'. The term includes all types of food or food products that have been produced (e.g. grown, manufactured, cooked) for human consumption and then thrown away. 'Food wastage' includes the following types of discarded food:

1.1.1 Food Losses

Food losses are wholesome edible material intended for human consumption that is instead lost as an unintended result of agricultural processes, lack of technology or technical limitations in storage, packaging, and/or marketing, poor infrastructure and logistics, insufficient skills, knowledge and management capacity of supply chain actors (FAO 2012a; Lipinski et al. 2013) or consumed by pests (Foresight 2010; Lin et al. 2013; Pfaltzgra et al. 2013). These take place all along food supply chain (FAO 2012b). Food losses may occur at the production, storage, processing, distribution, retail stages, as well as before, during or after meal preparation (BIO Intelligence Service et al. 2011).

1.1.2 Food Residuals

Food residuals are unavoidable inedible and partly avoidable wastes such as skins, bones, stalks, shells and leaves (WRAP 2009; Foresight 2010; BIO Intelligence Service et al. 2011). They also include residues generated in restaurants, pubs, coffee shops and certain food production facilities no longer intended for human consumption (Lin et al. 2013).

1.1.3 By-Products, Including Animal By-Products

By-product is a useful and marketable product or service deriving from a manufacturing process that is not the primary product or service being produced (EEA 2013). Food by-products are edible material that generated during food processing and manufacturing, and usually diverted away from the human food chain and fed to animals (Foresight 2010). Animal by-products are "entire bodies or parts of animals or products of animal origin … not intended for human consumption, including ova, embryos and semen" (European Parliament and Council of the European Union Regulation (EC) 1774/2002).

1.1.4 Food Waste

Food waste belongs to the category of avoidable waste. Discarded food still has value and is very often fit for consumption (FAO 2012a). Food waste is food that is spilled, spoiled, bruised or wilted. It may include whole or unopened packets or individual items of food which are not eaten at all (WRAP 2008). Food waste arises at any point in the food supply chain (Foresight 2010) as a result of inappropriate behaviour of food chain actors (e.g. producers, retailers, the food service sector, consumers) (FAO 2012b) as well as resulting from a lack of existing preventative technologies. A share of each of the aforementioned food wastage sub-types could be avoided by applying latest available instruments, therefore, until then, it might be considered as food waste.

Based on the analysis presented above and the hierarchy which exists, the current work will center on the 'food waste' sub-type.

References

BIO Intelligence Service, Umweltbundesamt, & Arcadis (2011). *Guidelines on the preparation of food waste prevention programmes*. Retrieved from http://ec.europa.eu/environment/waste/prevention/pdf/prevention_guidelines.pdf.

EEA (2013). By-product. *Environmental Terminology and Discovery Service (ETDS)*. Retrieved September 22, 2013, from http://glossary.eea.europa.eu/terminology/concept_html?term=by-product.

European Commission (2011). Communication from the Commission to the European Parliament, the Council, the European Economic and Social Committee and the Committee of the regions. *Roadmap to a Resource Efficient Europe COM(2011)571* (pp. 1–26). Retrieved from http://ec.europa.eu/environment/resource_efficiency/pdf/com2011_571.pdf.

European Parliament and Council of the European Union Regulation (EC) 1774/2002. *European Parliament and Council of the European Union Regulation (EC) No 1774/2002 of 3 October 2002 laying down health rules concerning animal by-products not intended for human consumption. Official Journal* L 273 10.10.2002. Retrieved from http://eur-lex.europa.eu/LexUriServ/LexUriServ.do?uri=OJ:L:2002:273:0001:0001:EN:PDF.

European Parliament Resolution (2011/2175(INI)) (2012). European Parliament resolution of 19 January 2012 on how to avoid food wastage: strategies for a more efficient food chain in the EU (2011/2175(INI)), 2175(January) (pp. 1–8). Retrieved from http://www.europarl.europa.eu/sides/getDoc.do?pubRef=-//EP//NONSGML+TA+P7-TA-2012-0014+0+DOC+PDF+V0//EN.

FAO (2012a). *Food wastage footprint. An environmental accounting of food loss and waste. Concept Note*. Retrieved from http://www.fao.org/fileadmin/templates/nr/sustainability_pathways/docs/Food_Wastage_Concept_Note_web.pdf.

FAO (2012b). *Global initiative on food losses and waste reduction*. Retrieved from http://www.fao.org/docrep/015/i2776e/i2776e00.pdf.

FAO (2013). *Food wastage footprint. Impacts on natural resources*. Retrieved from http://www.fao.org/docrep/018/i3347e/i3347e.pdf.

FAO, WFP, & IFAD (2012). *The State of Food Insecurity in the World 2012. Economic growth is necessary but not sufficient to accelerate reduction of hunger and malnutrition*, Rome. Retrieved from http://www.fao.org/docrep/016/i3027e/i3027e.pdf.

Foresight (2010). *How can waste reduction help to healthily and sustainably feed a future global population of nine billion people?, London*. Retrieved from http://www.bis.gov.uk/assets/foresight/docs/food-and-farming/workshops/11-608-w4-expert-forum-reduction-of-food-waste.pdf.

Foresight (2011). *The future of food and farming: Challenges and choices for global sustainability, London*. Retrieved from http://www.bis.gov.uk/assets/foresight/docs/food-and-farming/11-546-future-of-food-and-farming-report.pdf.

Gunders, D. (2012). *Wasted: How America is losing up to 40 percent of its food from farm to fork to landfill*. Retrieved from http://www.nrdc.org/food/files/wasted-food-ip.pdf.

Gustavsson, J., et al. (2011). *Global food losses and food waste—Extent, causes and prevention, Rome*. Retrieved from http://www.fao.org/docrep/014/mb060e/mb060e00.pdf.

Institution of Mechanical Engineers (2013). *Global food waste not, want not, London*. Retrieved from http://www.imeche.org/Libraries/Reports/Global_Food_Report.sflb.ashx.

Lin, C. S. K., et al. (2013). Food waste as a valuable resource for the production of chemicals, materials and fuels. Current situation and global perspective. *Energy & Environmental Science, 6*(2), 426–464. Retrieved August 13, 2013, from http://xlink.rsc.org/?DOI=c2ee23440h.

Lipinski, B., et al. (2013). *Reducing food loss and waste, Washington, DC*. Retrieved from http://www.worldresourcesreport.org.

Nellemann, C., et al. (2009). *The environmental food crisis—The environment's role in averting future food crises*. Retrieved from http://www.grida.no/files/publications/FoodCrisis_lores.pdf.

Pfaltzgra, L. A., et al. (2013). Food waste biomass: a resource for high-value chemicals. *Green Chemistry, 15*, 307–314. Retrieved from http://pubs.rsc.org/En/content/articlepdf/2013/gc/c2gc36978h.

U.S. EPA (2013). *2011 Municipal solid waste characterization report, Washington, DC*. Retrieved from http://www.epa.gov/waste/nonhaz/municipal/pubs/MSWcharacterization_fnl_060713_2_rpt.pdf.

Venkat, K. (2011). The climate change and economic impacts of food waste in the United States. *International Journal of Food System Dynamics, 2*(4), 431–446. Retrieved from http://www.cleanmetrics.com/pages/ClimateChangeImpactofUSFoodWaste.pdf.

Weber, B., Herrlein, S., & Hodge, G. (2011). *The challenge of food waste, London*. Retrieved from www.planetretail.net.

World Economic Forum (2009). *Driving sustainable consumption value chain waste driving sustainable consumption value chain waste* (pp. 1–8). Retrieved from http://www.weforum.org/pdf/sustainableconsumption/DSCOverviewBriefing-ValueChainWaste.pdf.

WRAP (2008). *The food we waste*. Retrieved from http://wrap.s3.amazonaws.com/the-food-we-waste.pdf.

WRAP (2009). *Household food and drink waste in the UK, Banbury*. Retrieved from http://www.wrap.org.uk/sites/files/wrap/HouseholdfoodanddrinkwasteintheUK-report.pdf.

Chapter 2
Literature Review

Every year 11.2 billion tonnes of solid waste are collected worldwide (UNEP 2011). In upcoming years the amount of accumulated waste will continue to increase together with growing population, an urbanization rate, overall economic and GDP/GNI per capita growth, an increase in production and consumption, and changes in a consumption pattern. Furthermore, the latest World Bank report predicts that annual global solid waste management costs will increase from USD 205.4 billion to about USD 375.5 billion by 2025 (Hoornweg and Bhada-Tata 2012). However, there is a positive aspect to this waste—its huge economic potential. Today the world waste market, from collection to recycling, is estimated at USD 410 billion a year, not including the sizable informal segment in developing countries (UNEP 2011).

According to the Eurostat data, the European Union alone generates about 3 billion tonnes of waste annually, and due to the OECD projections by 2020, this amount will increase by 45 % in comparison to 1995 (European Commission 2013b). Such a quantity of waste and its complexity not only have a significant adverse environmental impact, causing pollution, greenhouse gas emissions, and posing threats to human health, but also wastes a huge amount of material and energy resources (European Commission 2010; EEA 2013b).

Highly dependent on imported raw materials, Europe, in its long-term goals and strategies strives to reduce the amount of waste generated by improving its resource efficiency through recycling, avoiding waste and using unavoidable waste as a resource wherever it possible (European Commission 2010).

Waste prevention has been identified as one of the top priorities in the EU's Sixth Environment Action Programme (European Commission 2013b) as well as in the proposal of the European Commission for the 7th Environment Action Programme and the Roadmap to a resource efficient Europe (EEA 2013a).

The European Union's approach to waste management is based on the following principles:

- Waste prevention, which is closely linked with improving manufacturing methods and influencing consumers to demand greener products and less packaging.

W. Leal Filho and M. Kovaleva, *Food Waste and Sustainable Food Waste Management in the Baltic Sea Region*, Environmental Science and Engineering, DOI 10.1007/978-3-319-10906-0_2

- Recycling and reuse as an alternative to waste prevention in cases when it is not possible.
- Improvement of final disposal and monitoring as the last option, where waste is safely incinerated or landfilled (European Commission 2013b).

2.1 Legislation

These principles are reflected in the European framework of waste legislation. The framework includes a variety of requirements and technical standards for waste management in general (for all waste streams), for specific waste streams (e.g. packaging waste) and for specific waste treatment modes such as landfill and waste incineration (Neubauer 2007; EAUC 2013). All of these standards are implemented through a large number of EU Directives and Regulations, the cornerstone of which is the EU Waste Framework Directive considered as the "basic law" of the EU Waste Policy. The Directive dates from 1975 and was re-edited in 2006 (Neubauer 2007) as a result of the 2005 Thematic Strategy on Waste Prevention and Recycling (European Commission 2010).

The Thematic Strategy on the Prevention and Recycling of Waste (COM (2005) 666) adopted in 2005 (Commission of the European Communities Communication COM (2005) 666 2005) became a main driver for reforming out-dated principles and requirements of the EU waste legislation and bringing a new approach which is dictated by the realities of the world today. The Strategy defines the long term goal of switching the EU to a recycling society that seeks to avoid waste and uses waste as a resource. It promotes prevention, recycling and re-use measures as well as an application of a life-cycle orientated approach to waste management. It sets minimum EU standards for recycling activities and a framework for specific national policies. Moreover, the document recommends an improvement of the knowledge base on the impact of resource use, waste generation and management (Commission of the European Communities Communication COM (2005) 666 2005).

According to the Strategy, the Revised Waste Framework Directive (2008/98/EC) (2008) sees waste as a valued resource by strengthening its economic value and sets out targets for EU Member States to recycle 50 % of their municipal waste by 2020 (European Commission 2010). The countries are also required to introduce legislation on waste collection, reuse, recycling and disposal (European Commission 2013b). In addition to the definition of key concepts related to waste management, the document clarifies the difference between waste and by-products, sets criteria and conditions for situations when waste ceases to be waste and focuses on reducing the environmental impacts of waste generation. The Directive extends producer responsibilities and requires that the Member States establish waste management plans as well as waste prevention programs (Directive 2008/98/EC 2008).

However, based on the review of the progress towards achieving the Strategy's objectives, experts have stated that despite an improvement of legislation, increased recycling rates, a reduction of the amount of waste going to landfill and of the relative environmental impacts per ton of waste treated, after 5 years, the Strategy's main objectives still remain valid (European Commission Report COM (2011) 13 2011).

Another important directive that sets out the main requirements for waste disposal is the EU Landfill Directive (1999/31/EC) (Council Directive 1999/31/EC 1999). It is necessary to stress that by defining the term 'waste' the directive refers to the Council Directive on waste (75/442/EC) from 1975 (Council Directive 75/442/EEC 1975). The document includes a definition of waste types with no reference to the waste list adopted in Commission Decision 2000/532/EC (2000) (Commision Decision 2000/532/EC 2000), which could result in collisions, confusions, and a necessary revision of the Directive.

The Directive sets maximum capacities for landfill sites and defines targets for the reduction of biodegradable municipal waste (BMW) going to landfills. It also bans certain waste streams from being put into landfill sites. The document requires the member states to set up a national strategy for operations aimed at the reduction of BMW, such as recycling, composting, recovery and biogas production. It contains requirements for opening and maintaining a landfill during its operational and after-care phases (Council Directive 1999/31/EC 1999).

However, the results of the assessment of achievements in this area show that in 2010 despite significant successes in increasing material recycling the majority of the European countries still send more than half of their municipal waste to landfill (EEA 2013a).

The next significant document is the Directive (94/62/EC as amended by 2004/12/EC 2004) on packaging and packaging waste (European Parliament and Council Directive 94/62/EC 1994; Directive 2004/12/EC 2004), which takes precedence over the Waste Framework Directive where packaging and packaging waste are concerned (Arcadis et al. 2010). The document clarifies the definition of the term 'waste', by introducing a number of additional criteria and defines such operations as 'recovery', 'recycling', 'energy recovery', 'organic recycling' and 'disposal'. It also obliges the member states to set up return, collection and energy recovery systems, and to encourage the use of materials obtained from recycled packaging waste. A reduction of the overall volume of packaging is stated as the best means of preventing the creation of packaging waste. The document discusses a necessity of a harmonized reporting technique and clear guidelines for data provision. It also requires implementation of preventive measures with an emphasis on the minimization of environmental impact (Directive 2004/12/EC 2004).

2.2 Waste Management Hierarchy

Looking at food waste historically, The early 1970s could be considered as a turning point for waste management in Europe. The 1972 Report to the Club of Rome and the oil crisis in 1973 drew attention to an issue of the scarcity of raw materials. These events induced the change in societys' perception of the term 'waste', methods of waste handling and necessary transitions in waste management (Kemp and van Lente 2011). In 1979, a Dutch politician Ad Lansink developed a priority list for the various waste management methods, which became known as 'Lansink's Ladder' and became official policy in 1981 (Raven 2007). At the top of the Ladder is 'prevention of waste', followed by 're-use (of products)', 'recycling (of materials)', 'incineration (with energy-production)' and 'landfilling' as the last option (Kemp and van Lente 2011).

Today's waste prevention framework, which uses the 'Lansink's Ladder' as a prototype, is widely used in various waste related areas such as legislation and numerous projects, initiatives and strategies. The current framework is a five-step hierarchy of waste management and waste treatment options ordered according to what is best for the environment (UK Department of Energy and Climate Change and Defra 2011). It is a set of rules for waste management planning, qualified waste collection and treatment (Neubauer 2007). Such a framework is helpful for understanding how management approaches can be used to influence materials as they flow through the material life cycle (U.S. EPA 2009). However, in each particular case the hierarchy passes through "modifications". Having waste prevention as a final goal, different expert groups and institutions adjust the waste hierarchy by extending or narrowing the content of its stages.

In the US it is implemented by the U.S. Environmental Protection Agency (EPA). The EPA works under the Resource Conservation and Recovery Act, primary law, which governs the disposal of solid and hazardous waste in the country. Under this law the EPA encourages practices that reduce the amount of waste needing to be disposed of, such as waste prevention, recycling, and composting (U. S. EPA 2013b). The agency has ranked the most environmentally preferable options for waste management from 'source reduction' (including reuse) to 'treatment and disposal', with 'recycling', 'composting' and 'energy recovery' between (Fig. 2.1) (U.S. EPA 2012b).

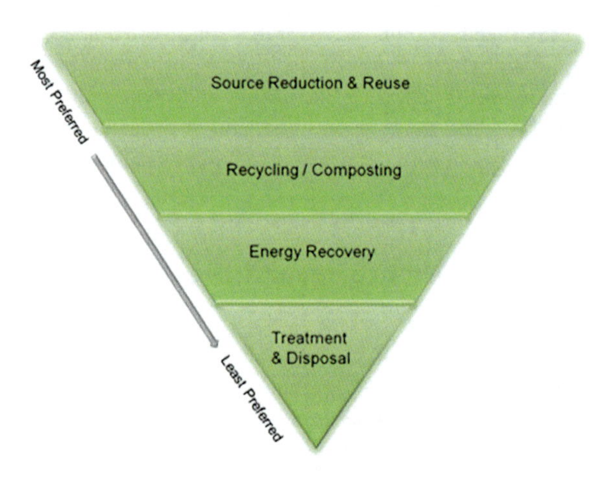

Fig. 2.1 Waste Management Hierarchy (U.S. EPA 2012b)

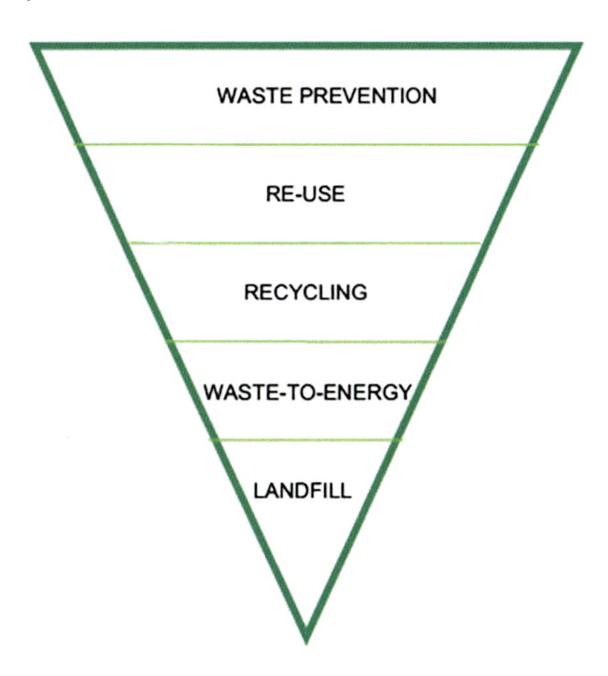

Fig. 2.2 Waste hierarchy (UNEP Division of Technology, Industry and Economics International Environmental Technology Centre 2010)

UNEP's various programmes and projects also endeavour to align with the waste management hierarchy (Fig. 2.2) used by the International Solid Waste Association (UNEP Division of Technology, Industry and Economics International Environmental Technology Centre 2010).

The association describes the hierarchy as 'a valuable conceptual and political prioritisation tool which can assist in developing waste management strategies aimed at limiting resource consumption and protecting the environment' (ISWA 2009).

A waste management hierarchy is also a framework used in the approach of Integrated Solid Waste Management (ISWM) (Fig. 2.3). This strategic concept is used for managing all sources of waste: prioritising waste avoidance and minimisation; practicing segregation; promoting the 3Rs (Reduce, Re-use, Recycle); implementing safe waste transportation; and treatment and disposal in an integrated manner with an emphasis on maximising resource-use efficiency (UNEP 2011).

Encouraged by the Thematic Strategy on the Prevention and Recycling of Waste (Commission of the European Communities Communication COM (2005) 666 2005), the EU waste policy has put an increasing focus on waste prevention (WRAP 2012). The waste management hierarchy and its stages (Fig. 2.4) are defined in the WFD.

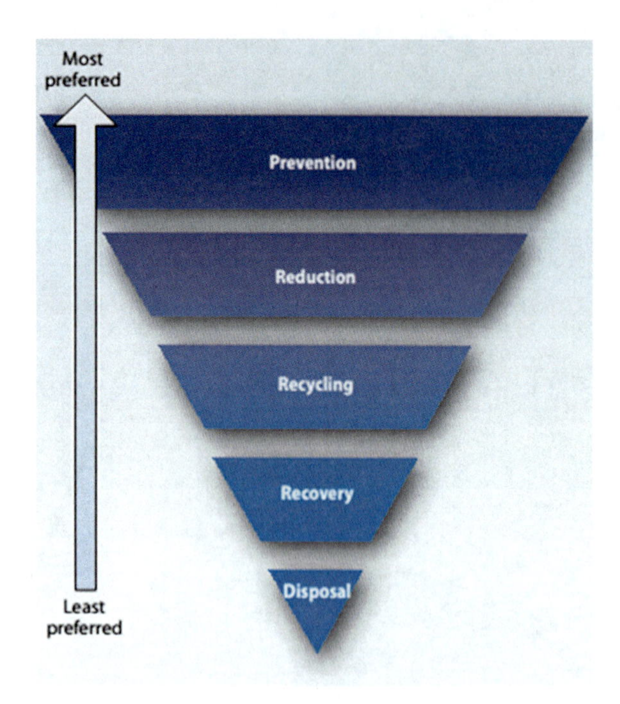

Fig. 2.3 The waste management hierarchy (UNEP 2011)

Fig. 2.4 The EU waste hierarchy (WRAP 2012)

2.2.1 Differences and Similarities in the Waste Management Hierarchies

All 'users' of the hierarchies agree upon the most and least preferable options for waste management. The top option 'waste prevention/avoidance/reduction' is stated as a crucial aspect of waste management (U.S. EPA 2012b; ISWA 2009; UNEP 2011; WRAP 2012; Directive 2008/98/EC 2008). However, institutions define this stage differently. The WFD defines 'prevention' as "measures taken before a substance, material or product has become waste that reduces:

- The quantity of waste, including through the re-use of products or the extension of the life span of products;
- The adverse impacts of the generated waste on the environment and human health; or
- The content of harmful substances in materials and products".

(Directive 2008/98/EC 2008).

In its definition, the EPA also stresses such waste prevention techniques as donating items, buying in bulk and reducing packaging (U.S. EPA 2012b). At the same time the ISWM approach distinguishes between 'prevention' and 'reduction' stages (Fig. 2.3) (UNEP 2011).

The results achieved during this stage of the hierarchy are very important because it leads to resource conservation (WRAP 2012) and eliminates the need to dispose something that is not produced. Yet this is a very challenging concept because it is difficult to measure something which, by definition, never existed (European Commission 2010).

The next step, further down the WFD's hierarchy, is 'preparing for re-use' (Fig. 2.4). The directive differentiates between 'preparing for re-use', which means checking, cleaning or repairing, recovery operations, by which products or components of products that have become waste are prepared so that they can be re-used without any other pre-processing, and 're-use', which means any operation by which products or components that are not waste are used again for the same purpose for which they were conceived (Directive 2008/98/EC 2008). This is one of the main differences between the WFD's waste management hierarchy and the frameworks used by other institutions. At the same time the EPA merges the 'source reduction' and 're-use' stages (Fig. 2.1), whereas the waste management hierarchy, used in the ISWM approach, does not include the 're-use' option as a separate/stand-alone stage in general (Fig. 2.3) (UNEP 2011).

'Recycling' means a series of activities that includes collecting recyclable materials that would otherwise be considered waste, sorting and reprocessing into products, materials or substances whether for the original or other purposes (Directive 2008/98/EC 2008).

In the EPA's version of the hierarchy 'recycling' and 'composting' stages are merged (Fig. 2.1). Moreover, the Agency considers composting of food scraps, yard trimmings, and other organic materials as a part of the 'recycling' options. The definition again points out the importance of consumers who provide the last link in recycling by purchasing products made from recycled content (U.S. EPA 2012b).

The next step in every version of the hierarchy is 'energy recovery' from waste. The EPA defines it as "the conversion of non-recyclable waste materials into useable heat, electricity, or fuel through a variety of processes, including combustion, gasification, anaerobic digestion, and landfill gas recovery" (U.S. EPA 2012b).

The WFD sees 'energy recovery' as one of many recovery options. It defines recovery as "any operation the principal result of which is waste serving a useful purpose by replacing other materials which would otherwise have been used to fulfil a particular function, or waste being prepared to fulfil that function, in the plant or in the wider economy." It provides a list of recovery operations, which among others includes such operations as recycling/reclamation of metals and metal compounds, regeneration of acids or bases, oil re-refining or other reuses of oil, land treatment resulting in benefit to agriculture or ecological improvement and etc. (Directive 2008/98/EC 2008).

The last and least preferable option, which all actors agree upon, is 'disposal'. This stage includes landfilling and incineration without energy recovery. The WFD defines 'disposal' as "any operation which is not recovery even where the operation has as a secondary consequence the reclamation of substances or energy" (Directive 2008/98/EC 2008). In addition to these options the EPA adds collection and usage of methane as fuel to generate electricity and includes future possibilities of usage of capped landfills as recreation sites such as parks, golf courses, and ski slopes (U. S. EPA 2012b).

In addition, it is worth remarking that in a certain case if some options are not stressed as separate stages, it does not mean that these have not been considered by experts. Such a situation might be perceived as a way to leave more space and flexibility for their activities in the frame of this concept.

One of the main purposes of the EU waste legislation is to move up the waste management hierarchy (European Commission 2010). However, according to the Eurostat data the main methods of waste treatment in EU-28 in 2010 were 'recovery other than energy recovery', 'disposal' and 'deposit onto or into land' (Fig. 2.5). Even despite the fact that sending waste to landfill is considered as the worst waste management option it is still one of the most used municipal solid waste (MSW) disposal methods in the EU (Commission of the European Communities Green Paper COM (2008) 811 2008; European Commission 2010, 2012).
Tagore 2011 has been changed to Tagore 2010 so that this citation matches the list.

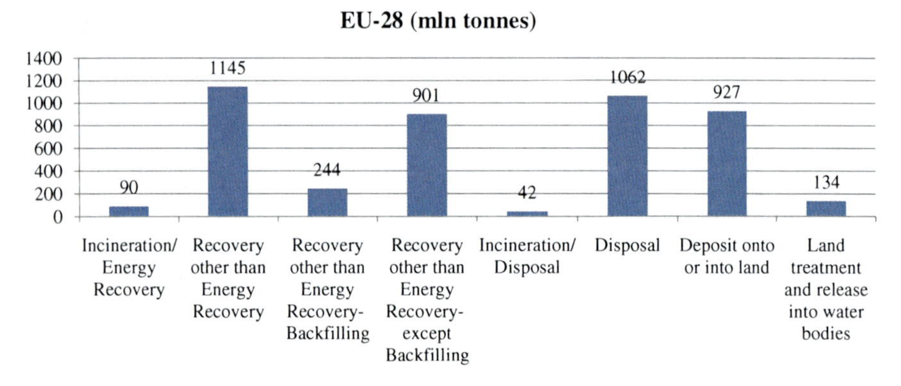

Fig. 2.5 Treatment of waste in EU-28 in 2010, in million tonnes (Eurostat 2013a, b)

On a national level, the WFD obliges countries to develop national waste management plans which include a baseline analysis of the current waste management situation in that country. Furthermore, the MS are required to establish National Waste Prevention Programmes by the end of 2013, which clearly identify the waste prevention measures and objectives (Directive 2008/98/EC 2008). In order to support the MS during the development of these programmes, the European Commission [DG Environment] has prepared a guidance document (European Commission Directorate-General Environment 2012). The guide provides detailed information about the stages of the waste hierarchy, and relevant EU waste prevention strategies and initiatives. It defines key waste streams, key stakeholders and types of waste that they produce. In addition, the document offers a procedure for planning and implementing a national waste prevention programme, and lists principle approaches which identify the most efficient measures for it (European Commission Directorate-General Environment 2012).

2.3 Bio-Waste

One of these waste types, which draw particular attention in the waste management policies, is bio-waste > This is mainly because of the environmental threats associated with its decomposition in landfills. The amount of bio-waste accounts for about one third of the waste generated by households in the EU. Each year Europe produces between 118 and 138 million tonnes of bio-waste, of which about 88 million tonnes is municipal waste (European Commission Communication COM (2010) 235 2010), on average, 40 % of this type of waste goes to landfill (European Commission 2010). Annually, the decay of the organic proportion of solid waste is contributing to about 5 % of global Greenhouse Gas (GHG) emissions (UNEP 2011). Experts talk about a need for greater focus on bio-waste recycling in line with the waste hierarchy (EEA 2013a). Arcadis et al. 2010, based on a multi-criteria-assessment, prioritized the bio-waste flow as one of the top four priority material and waste flows which have to become target areas for waste prevention measures (Arcadis et al. 2010). In addition the experts predict an increase in the share of bio-waste in the total generation of MSW at the EU-27 level, which will reach about 35.6 % by year 2020 (Fig. 2.6).

The WFD defines 'bio-waste' as "biodegradable garden and park waste, food and kitchen waste from households, restaurants, caterers and retail premises and comparable waste from food processing plants" (Directive 2008/98/EC 2008). It does not cover forestry or agricultural residue and it should not be confused with the wider term "biodegradable waste" (European Commission Communication COM (2010) 235 2010). The Landfill Directive defines 'biodegradable waste' as "any waste that is capable of undergoing anaerobic or aerobic decomposition, such as food and garden waste, and paper and paperboard" (Council Directive 1999/31/EC 1999). Thus bio-waste excludes paper and paperboard waste.

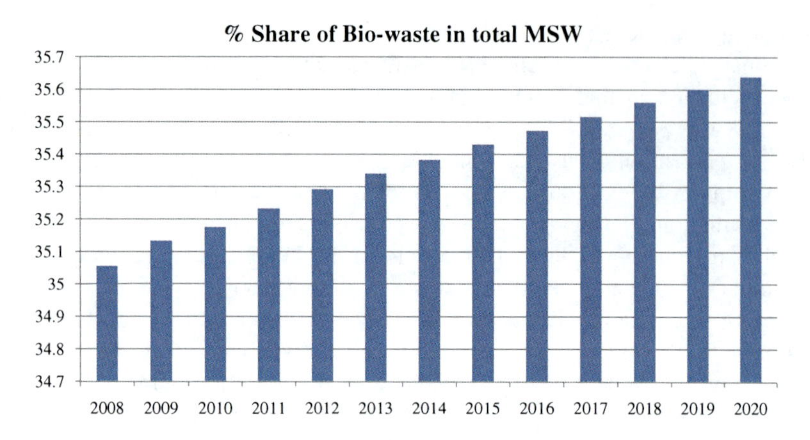

Fig. 2.6 The projection of % share of Bio-waste in total MSW generation at the EU-27 level (Arcadis 2010)

Today, the set of available techniques for the bio-waste treatment includes prevention at source, separate collection, biological treatment such as anaerobic digestion and composting, incineration, and landfill (Commission of the European Communities Green Paper COM (2008) 811 2008). Landfill and incineration methods are prevailing (Fig. 2.7) because they are considered by these countries as one of the easiest and cheapest options for bio-waste treatment (European Commission 2012). Aiming to change the current situation and move up the waste hierarchy the European Commission has taken a number of steps reflected in the EU waste policy and legislation.

In order to reduce the amount of bio-waste sent to landfill, the Landfill Directive sets binding targets regarding the allowed amount of municipal biodegradable waste to be landfilled. It should be reduced to 35 % of 1995 levels by 2016, which

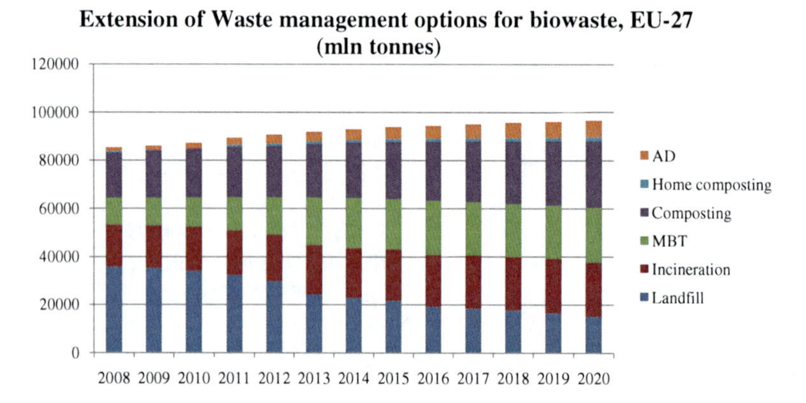

Fig. 2.7 The projection of extension of Bio-waste management options at the EU-27 level (Arcadis 2010)

leads to significant reduction of the problem of methane production (European Commission 2010). The 'dead-line' for such countries as Estonia, Latvia, Lithuania etc. is extended to 2020 (European Commission Directorate-General Environment 2012).

To discourage the incineration of bio-waste with low calorific value, the WFD defines energy efficiency levels below which the incineration of municipal solid waste may not be regarded as recovery (European Commission Communication COM (2010) 235 2010). The incineration of bio-waste is regulated by the Waste Incineration Directive (Directive 2000/76/EC 2000), which sets emission limit values and monitoring requirements for pollutants to air (Arcadis 2010; Commission of the European Communities Green Paper COM (2008) 811 2008). The WFD also requires the MS to encourage the use of materials produced from bio-waste, and to consider future options of bio-waste composting and digestion through separate collection (Directive 2008/98/EC 2008). The benefits of separate collection include easy diversion of biodegradable waste from landfills, enhancement of the calorific value of the remaining MSW, and generation of a cleaner bio-waste fraction that allows to produce high quality compost and facilitates biogas production (Commission of the European Communities Green Paper COM (2008) 811 2008; European Commission 2010, 2012). Results of the EC study on the 'evolution of (bio-) waste generation/prevention and (bio-) waste prevention indicators' (Reisinger et al. 2011) showed that some MS (or regions of the MS) have already implemented programmes for separate collection, diversion from landfill of bio-waste, and prevention of bio-waste via the use of economic instruments or targets. However, no national prevention targets specific to bio-waste were identified (Reisinger et al. 2011). At the same time a Green Paper, which explored options for the further development of the bio-waste management in the EU (BIO Intelligence Service 2010), published by the Commission in 2008, stated that there are no easy administrative solutions for bio-waste prevention and possible actions should be generally linked to changing consumer behaviour and retail policies (Commission of the European Communities Green Paper COM (2008) 811 2008).

Experts have also identified additional obstacles associated with the implementation of alternative methods to landfill to treat bio-waste. Firstly, bio-waste management options depend on a variety of factors such as inter alia collection systems, waste composition and quality, climatic conditions, population density, and the potential of use of various waste-derived products such as electricity, heat, methane-rich gas or compost (Commission of the European Communities Green Paper COM (2008) 811 2008; European Commission 2012), which define their environmental and economic benefits. Therefore, the EU legislation does not limit Member States' choices of bio-waste treatment options. The choice of treatment options needs to be explained and justified in national or regional Waste Management Plans and Prevention Programmes (Commission of the European Communities Green Paper COM (2008) 811 2008). Secondly, the results of the EC assessment of feasibility of setting bio-waste recycling targets in the EU outlined the following barriers of the implementation of separate collection and recycling of bio-waste:

- A general lack of experience and knowledge of the benefits of recycling/separate collection, the methods to set up a successful collection scheme, the cost structures, the ways to ensure compost/digestate quality, the uses of compost/digestate, the market functioning of waste-derived products such as compost, etc.;
- The costs linked to separate collection and recycling;
- Political barriers, logistical and social issues, mainly in rural areas and city centres.

(Vito et al. 2011).

To overcome these obstacles and to assist decision-makers in making the best use of biodegradable waste in line with the waste hierarchy, the Commission recommends to use the Life Cycle Assessment tool and Life Cycle Thinking approach to plan the management of bio-waste (European Commission 2012). Such an approach can be used alongside the waste hierarchy in order to make sure that the best overall environmental option is identified (European Commission et al. 2011). The Green Paper recommends the Commission to provide additional measures to support incineration with high energy recovery, anaerobic digestion with biogas production and recycling of bio-waste (Commission of the European Communities Green Paper COM (2008) 811 2008). Moreover, the production of good quality compost and bio-gas contributes to enhanced soil quality and resource efficiency, as well as a higher level of energy self-sufficiency (European Commission 2012).

Despite aforementioned constraints and barriers, efforts undertaken by the Commission such as legal restrictions and the support of a variety of programmes, projects and initiatives are having a positive effect. Experts, in their projection of the extension of bio-waste management options to 2020, predict an increase in the extension of MBT together with other treatment methods which will lead to a significant decrease in usage of the option of landfill (Fig. 2.7) (IEEP et al. 2010; Arcadis 2010).

Aligning the management of bio-waste with the waste hierarchy and other provisions of the WFD could have both financial and environmental benefits. Due to the communication from the Commission to the Council and the European Parliament on future steps in bio-waste management in the EU, the financial benefits could range between EUR 1.5–EUR 7 billion depending on the extent of recycling, and as a result environmental benefits could include an approximately 34 million tonnes CO_2 equivalent saving, 80–90 % of which would be due to prevention (European Commission Communication COM (2010) 235 2010).

2.4 Food Waste

However, the type of bio-waste that raises the biggest concern is food waste, which is the main focus of the current study. The problem of food waste takes a very particular place, by touching not only such issues as depletion of natural resources, environmental pollution and climate change, but also ethical and social aspects of throwing away food, where due to the FAO estimations about 870 million people globally were suffering from chronic undernourishment in 2010–2012 (FAO 2013a). According to the FAO, about a third (around 1.3 billion tonnes) of the food for human consumption is wasted globally (FAO 2013b) and about 90 million tonnes of food is wasted annually in Europe (European Commission 2013a), where 16 million citizens receive food aid from charitable institutions (European Parliament Resolution (2011/2175 (INI)) 2012).

Food waste is responsible for various negative environmental effects with the high relevant costs (Bakas and Herczeg 2010). Food loss and waste impact on global climate change, resulting in unnecessary greenhouse gas emissions and inefficiently used water and land. This in turn leads to diminished natural ecosystems and the services they provide (Hall et al. 2009; Foresight 2011; Lipinski et al. 2013).

Economically, food loss and waste amounts to roughly USD 680 billion in industrialized countries and USD 310 billion in developing countries (FAO 2013b). In addition, it represents wasted investments that can reduce producers' incomes and increase consumers' expenses (Lipinski et al. 2013). In addition, ethically, it results in missed opportunities to feed the growing world population (FAO 2012). One quarter of the food currently lost or wasted globally could be saved, and this would be enough to feed the 870 million people globally who are in need of food (FAO 2013b).

Food waste has been identified by the European Commission as the most important household waste stream that must be prevented, therefore, the reduction of food waste required to be the core of any biodegradable waste (or bio-waste) prevention activity (European Commission Directorate-General Environment 2012), and the support of such activities on the EU level, would have the biggest impact (Reisinger et al. 2011).

Oddly enough, the definition of the term 'food waste' arises in many discussions as the problem of its generation. The interpretation of the term depends on each particular research group and the boundaries of group's work.

In 1981 the FAO defined 'food' as weight of whole some edible material that would normally be consumed by humans in the book "Food loss prevention in perishable crops". Inedible portions such as skins, stalks, leaves, and seeds, potential foods (e.g., leaf protein), feed (intended for consumption by animals) were not defined as food. 'Loss' was defined as any change in the availability, edibility, wholesomeness or quality of the food that prevents it from being consumed by people (FAO 1983).

Based on the aforementioned definition Escaler and Teng 2011 define 'food loss or waste' as "edible material intended for human consumption, arising at any point in the food supply chain that is instead discarded, lost, degraded or consumed by pests between harvest and reaching the consumer" (Escaler and Teng 2011).

The FAO in its report "Global losses and Food waste" uses the following definition (Parfitt et al. 2010): "food losses occurring at the end of the food chain (retail and final consumption), which relates to retailers' and consumers' behaviour"

(Gustavsson et al. 2011). The experts do not make a clear differentiation between food losses and food waste by defining it as "the masses of food lost or wasted in the part of food chains leading to "edible products going to human consumption" (Gustavsson et al. 2011).

At the beginning of 2012 the European Parliament released a resolution on "how to avoid food wastage: strategies for a more efficient food chain in the EU", where 'food waste' was defined as "all the foodstuffs discarded from the food supply chain for economic or aesthetic reasons or owing to the nearness of the 'use by' date, but which are still perfectly edible and fit for human consumption and, in the absence of any alternative use, are ultimately eliminated and disposed of, generating negative externalities from an environmental point of view, economic costs and a loss of revenue for businesses" (European Parliament Resolution (2011/2175 (INI)) 2012).

Lipinski et al. 2013 defines 'food waste' as "food that is of good quality and fit for human consumption but that does not get consumed because it is discarded—either before or after it spoils" (Lipinski et al. 2013).

Such variety of different views requires inclusion of additional criteria to characterize food waste. The first one is food waste classification.

WRAP classified food waste into three types due to an availability rating:

- **avoidable food waste**—the food has been thrown away because it is no longer wanted or has been allowed to go past its best (e.g. an apple, half a pack of cheese, milk, or an fruit juice);
- **possibly avoidable food waste**—this is food that some people will eat and others will not, or that can be eaten when prepared in one way but not in another (e.g. bread crusts);
- **unavoidable food waste**—this waste arises from food preparation and includes foods such as meat bones and hard vegetable or fruit peelings (e.g. melon rind) (WRAP 2008), which have not been considered as food by FAO in 1981 in the first place.

Avoidable food waste gives rise to the biggest concern because it is food that could have been eaten if it had been better managed or stored. The food may not have been fit for consumption at the time of disposal because it had gone mouldy or had been spoilt (WRAP 2008; NSW Office of Environment and Heritage Australia 2011). The resolution on "how to avoid food wastage: strategies for a more efficient food chain in the EU" addresses main aspects regarding the problem of food waste and lists causes of food waste such as overproduction, faulty product targeting (unadapted size or shape), deterioration of the product or its packaging, marketing rules (problems of appearance or defective packaging), and inadequate stock management or marketing strategies (European Parliament Resolution (2011/2175 (INI)) 2012).

Another question to discuss is about the stage in a food supply chain where food becomes food waste/loss. Food losses and waste occur along a food supply chain in both developed and developing countries (World Economic Forum 2009). In developing countries over 40 % (European Commission 2013a) of food loss/waste arise at production, harvest, processing, storage and transportation stages, whereas

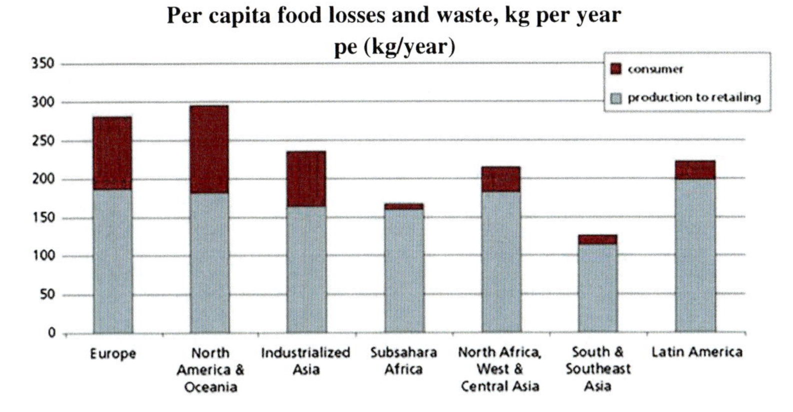

Fig. 2.8 Per capita food losses and waste, at consumption and pre-consumptions stages, in different regions (FAO 2013b)

in developed countries the majority of food waste is generated at the retail and consumption stages (Foresight 2011). In Europe and North America waste per capita by consumers is between 95–115 kg a year, while in sub-Saharan Africa, south and south-eastern Asia this amount does not exceed 6–11 kg a year (Fig. 2.8) (FAO 2013b).

Until today, a few frameworks have been suggested for food waste treatment. Through an analogy with the waste management hierarchy, the EPA developed 'Food waste recovery hierarchy' (Fig. 2.9) (U.S. EPA 2013a).

At the top of the hierarchy 'Source Reduction', followed by 'Feed Hungry People' which includes donation of extra food to food banks, recovery programs, soup kitchens, and shelters, 'Food Animals'—diversion of food scraps to animal

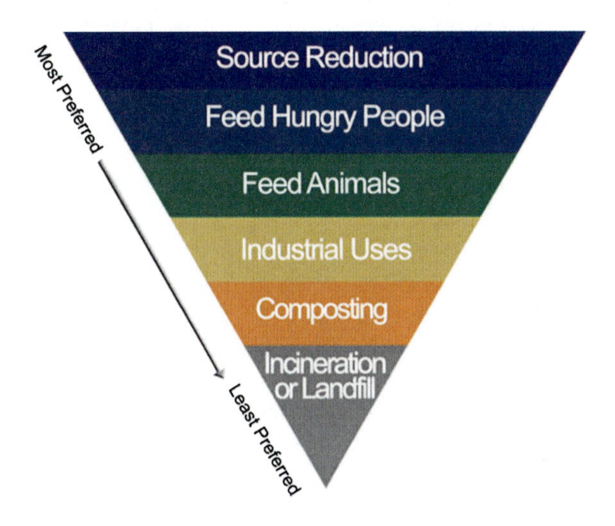

Fig. 2.9 U.S. EPA Food waste recovery hierarchy (U. S. EPA 2013a)

feed, 'Industrial Uses'—provision of waste oils for rendering and fuel conversion, and food scraps for digestion to recover energy, 'Composting'—creation of a nutrient-rich soil amendment, and the least preferable option is 'Landfill/Incineration' (U.S. EPA 2012a).

In the Netherlands, the Dutch Ministry of Economic Affairs, Agriculture and Innovation (EL&J), in parallel with Lansink's Ladder, uses the name 'Moerman Ladder'. Moerman's Ladder shows the 'optimum utilisation' of residual flows based on an ethical norm, prompted by worldwide food security problems (Waarts et al. 2011). The Ladder begins with the 'prevention' stage (avoiding food waste), where optimum use is food. The rest of the stages are

- 'use for human food' (for example food banks, Salvation Army);
- 'conversion to human food' (processing and reprocessing);
- 'use in animal feed';
- 'raw materials for industry' (biobased economy);
- 'processing to make fertiliser for cofermentation' (and energy generation);
- 'processing to make fertiliser through composting';
- 'use for sustainable energy' (objective is energy generation);
- 'burning as waste' (objective is destruction, with associated possibility of energy generation);
- 'dumping'

(Waarts et al. 2011).

The EC study on the "evolution of (bio-) waste generation/prevention and (bio-) waste prevention indicators" (Reisinger et al. 2011) compiled a list of recommended measures for the EU-action plan for food waste prevention, which includes but is not limited to:

- The setting of EU food waste reporting requirements (food waste generation, food waste in household waste);
- The setting of information/labelling requirements on resource efficiency and hazardous substance concentration of food (taking into account the potential of mobile and internet technology based information dissemination);
- The dissemination of best practice on more efficient food production and use of food, including logistical improvements (e.g. stock management improvements for retailers, reservation requirements for cafeterias, ordering flexibility in hospitals);
- The responsibility of waste prevention concepts and train planners to produce waste prevention concepts;
- Help organising networks on the redistribution of food;
- The clarification of standards (e.g. for setting "use by" and "best before" dates) by taking into consideration food safety;
- Awareness/information/motivation campaigns on food waste prevention;
- The tools and training for more efficient consumption (residual food cook books, shopping list and etc.)

(Reisinger et al. 2011).

At the European level, a list of necessary measures is stated in the resolution of the European Parliament. The document, amongst other things, calls as a matter of urgency the problem of food waste along the entire supply and consumption chain to be addressed, and to devise guidelines for and support ways of improving the efficiency of the food supply chain sector by sector, as well as to analyse the causes and effects of the disposal, wastage and landfilling, and associated economic, environmental, nutritional and social impacts. It asks to take practical measures towards halving food waste by 2025 and create specific food waste prevention targets for the Member States, as a part of the waste prevention targets to be reached by Member States by 2014 (European Parliament Resolution (2011/2175 (INI)) 2012). However in 2011 in the communication to the European Parliament "Roadmap to a Resource Efficient Europe" the Commission has set out the same target of a 50 % reduction of the disposal of food waste by the year 2020 (European Commission 2011). The communication has also pointed out that the wide-spreading of incentives to healthier and more sustainable food production and consumption would lead to a 20 % reduction in the food chain's resource inputs (European Commission 2011).

In order to support policymakers in developing National Waste Prevention Programmes (as well as waste management organisations, businesses, institutions, local authorities and environmental protection agencies and other actors dealing with food waste) the European Commission has prepared specific guidelines to address food waste. The "guidelines on the preparation of food waste prevention programmes," provides a general approach to food waste prevention, guidelines for developing a food waste prevention programme and addresses such key sectors as local authorities, households, the hospitality industry, the retail supply chain, businesses and institutions (such as schools and hospitals) (BIO Intelligence Service, Umweltbundesamt and Arcadis 2011).

In addition, the problem of food waste partly or entirely is covered in the following legal document: the Thematic Strategy on the Prevention and Recycling of Waste, the Landfill and Revised Waste Framework Directives, the Integrated Pollution Prevention and Control Directive, Incineration Directive, Nitrate Directive and EU Policy for Renewable Energy, Regulation on Animal By-Products, which constitutes the cornerstone of European legislation on food safety (Bakas and Herczeg 2010; Arcadis 2010).

It is important at this point to highlight that in the EU food waste is perceived as bio-waste and therefore all measures are applied from this perspective.

The most recent document "A Communication on Sustainable Food," is planned to be adopted by the EU in 2013 (EU-FUSIONS 2012).

Nevertheless, despite steps that have been taken, there is still a gap in data and information regarding the state of the problem of food waste in the MS. Moreover, a number of systematic investigations across the Europe regarding causes of food waste and ways of its reduction are very small. The most significant studies have been undertaken on behalf of the UK body, the Waste and Resources Action

Programme (WRAP) (Arcadis 2010) and by BIO Intelligence Service the 'Preparatory study on food waste across EU 27' (BIO Intelligence Service 2010).

Thereby, summing up the aforementioned points the following conclusions can be drawn:

- There is no single definition of the term 'food waste' as well as no clear classification and differentiation between what can be or cannot be considered as food waste.
- On a European level, there is some recognition of the problem of food waste and an acknowledged necessity to take measures to address it.
- One of the first steps is to reduce an existing gap in data and information about the state of the problem of food waste in the European countries.
- Currently, the number of conducted studies regarding volumes of food thrown away, its types, causes and applied methods of food waste reduction is very limited.

References

Arcadis (2010). *Assessment of the options to improve the management of bio-waste in the European Union—final report*. Retrieved from http://ec.europa.eu/environment/waste/compost/pdf/ia_biowaste—finalreport.pdf.

Arcadis et al. (2010). *Analysis of the evolution of waste reduction and the scope of waste prevention. Final Report*. Retrieved from http://ec.europa.eu/environment/waste/prevention/pdf/report_waste.pdf.

Bakas, B. I., & Herczeg, M. (2010, September). Food Waste. *CORPUS The SCR Knowledge Hub* (pp. 1–4). Retrieved from http://www.scpknowledge.eu/sites/default/files/BakasandHerczeg2010FoodWaste.pdf.

BIO Intelligence Service (2010). *Preparatory study on food waste across EU 27*, Retrieved from http://ec.europa.eu/environment/eussd/pdf/bio_foodwaste_report.pdf.

BIO Intelligence Service, Umweltbundesamt & Arcadis (2011). *Guidelines on the preparation of food waste prevention programmes*, Retrieved from http://ec.europa.eu/environment/waste/prevention/pdf/prevention_guidelines.pdf.

Commision Decision 2000/532/EC (2000). *Commision Decision 2000/532/EC of 3 May 2000 replacing Decision 94/3/EC establishing a list of wastes pursuant to Article 1(a) of Council Directive 75/442/EEC on waste and Council Decision 94/904/EC establishing a list of hazardous waste pursuant to Artic*, Official Journal of the European Communities. Retrieved from http://eurlex.europa.eu/LexUriServ/LexUriServ.do?uri=OJ:L:2000:226:0003:0003:EN:PDF.

Commission of the European Communities Communication COM (2005). 666 (2005). *Communication from the commission to the council, the european parliament, the European economic and social committee and the committee of the regions COM (2005) 666 Taking sustainable use of resources forward: A thematic strategy on the prevention and re (pp. 1–13)*. Retrieved from http://eurlex.europa.eu/LexUriServ/LexUriServ.do?uri=COM:2005:0666:FIN:EN:PDF.

Commission of the European Communities Green Paper COM (2008). 811, 2008. *Commission of the European Communities Green Paper COM (2008) 811 on the management of bio-waste in the European Union*, Brussels. Retrieved from http://eurlex.europa.eu/LexUriServ/LexUriServ.do?uri=COM:2008:0811:FIN:EN:PDF.

Council Directive 1999/31/EC (1999). *Council Directive 1999/31/EC of 26 April 1999 on the landfill of waste*, Official Journal L 182, 16/7/1999 (pp. 1–19). Retrieved from http://eurlex.europa.eu/LexUriServ/LexUriServ.do?uri=OJ:L:1999:182:0001:0019:EN:PDF.

Council Directive 75/442/EEC (1975). *Council Directive 75/442/EEC of 15 July 1975 on waste*, Official Journal L 194, 25/7/1975 (p. 39). Retrieved from http://eurlex.europa.eu/LexUriServ/LexUriServ.do?uri=CONSLEG:1975L0442:20031120:EN:PDF.

Directive 2000/76/EC (2000). *Directive 2000/76/EC of the European parliament and of the Council of 4 December 2000 on the incineration of waste*, Official Journal L 332, 28/12/2000 (pp. 91–111). Retrieved from http://eurlex.europa.eu/LexUriServ/LexUriServ.do?uri=OJ:L:2000:332:0091:0111:EN:PDF.

Directive 2004/12/EC (2004). *Directive 2004/12/EC of the European parliament and of the council of 11 February 2004 amending Directive 94/62/EC on packaging and packaging waste*, Official Journal L 47, 18/2/2004. Retrieved from http://eurlex.europa.eu/LexUriServ/LexUriServ.do?uri=OJ:L:2004:047:0026:0031:EN:PDF.

Directive 2008/98/EC (2008). *Directive 2008/98/EC of the European parliament and of the council of 19 November 2008 on waste and repealing certain directives*, European Union: Official Journal L 312, 22/11/2008. Retrieved from http://eurlex.europa.eu/LexUriServ/LexUriServ.do?uri=OJ:L:2008:312:0003:0030:en:PDF.

EAUC (2013). *EU waste strategy. Background to waste management* Retrieved August 19, 2013, fromhttp://www.eauc.org.uk/eu_waste_strategy .

EEA (2013a). *Managing municipal solid waste—a review of achievements in 32 European countries*, Copenhagen. Retrieved from http://www.eea.europa.eu/publications/managing-municipal-solid-waste.

EEA (2013b). *Municipal waste generation*. Retrieved August 26, 2013 from http://www.eea.europa.eu/data-and-maps/indicators/municipal-waste-generation.

Escaler, M., & Teng, P. (2011). "Mind the Gap": Reducing waste and losses in the food supply chain food losses in the food supply chain. *The Centre for Non-Traditional Security (NTS) Studies, S. Rajaratnam School of International Studies (RSIS), NTS Insight*, (6). Retrieved from http://www.rsis.edu.sg/nts/HTMLNewsletter/Insight/pdf/NTS_Insight_jun_1101.pdf/

EU-FUSIONS (2012). *About food waste*. Retrieved August 17, 2013 from http://www.eu-fusions.org/about-food-waste.

European Commission (2010). *Being wise with waste: The EU 's approach to waste management*, pp. 1–16. Retrieved from http://ec.europa.eu/environment/waste/pdf/WASTEBROCHURE.pdf.

European Commission (2011). *Communication from the commission to the European parliament, the Council, the European economic and social committee and the committee of the regions.* Roadmap to a Resource Efficient Europe COM(2011) 571. pp. 1–26. Retrieved from http://ec.europa.eu/environment/resource_efficiency/pdf/com2011_571.pdf.

European Commission (2012). *Biodegradable waste*. Retrieved March 25, 2013 from http://ec.europa.eu/environment/waste/compost/index.htm.

European Commission (2013a). *Health and consumers, food*. Retrieved August 17, 2013 from http://ec.europa.eu/food/food/sustainability/.

European Commission (2013b). Waste. *Environment*. Retrieved August 19, 2013 from http://ec.europa.eu/environment/waste/.

European Commission Communication COM (2010). 235 (2010). *Communication from the Commission to the Council and the European Parliament on future steps in bio-waste management in the European Union COM(2010)235 (pp. 1–12)*. Retrieved from http://ec.europa.eu/environment/waste/compost/pdf/com_biowaste.pdf.

European Commission Directorate-General Environment (2012). *Waste prevention—handbook: Guidelines on waste prevention programmes, (pp. 1–62)*. Retrieved from http://ec.europa.eu/environment/waste/prevention/pdf/Wastepreventionguidelines.pdf.

European Commission, Joint Research Centre & Institute for Environment and Sustainability (2011). *Supporting environmentally sound decisions for bio-waste management, (pp. 1–106)*. Retrieved from http://lct.jrc.ec.europa.eu/pdf-directory/D4A-Guidance-on-LCT-LCA-applied-to-BIO-WASTE-Management-Final-ONLINE.pdf.

European Commission Report COM (2011). 13, 2011. *European Commission Report to the European Parliament, the Council, the European Economic and Social Committee and the Committee of the Regions on the thematic strategy on the prevention and recycling of waste*, Brussels. Retrieved from http://eurlex.europa.eu/LexUriServ/LexUriServ.do?uri=COM:2011:0013:FIN:EN:PDF.

European Parliament and Council Directive 94/62/EC (1994). *European Parliament and Council Directive 94/62/EC of 20 December 1994 on packaging and packaging waste, Official Journal L 365, 31/12/1994 pp. 10–23*. Retrieved fromhttp://eur-lex.europa.eu/LexUriServ/LexUriServ.do?uri=OJ:L:1994:365:0010:0023:EN:PDF.

European Parliament Resolution (2011/2175 (INI)) (2012). *European Parliament Resolution of 19 January 2012 on how to avoid food wastage: Strategies for a more efficient food chain in the EU (2011/2175(INI)), 2175(January) (pp. 1–8)*. Retrieved from http://www.europarl.europa.eu/sides/getDoc.do?pubRef=-//EP//NONSGML+TA+P7-TA-2012-0014+0+DOC+PDF+V0//EN.

Eurostat (2013a). *Children at risk of poverty or social exclusion*. Retrieved December 22, 2013 from http://epp.eurostat.ec.europa.eu/statistics_explained/index.php/Children_at_risk_of_poverty_or_social_exclusion.

Eurostat (2013b). *Environment in the EU27 (March) (pp.1–3)*. Retrieved from http://epp.eurostat.ec.europa.eu/cache/ITY_PUBLIC/8-04032013-BP/EN/8-04032013-BP-EN.PDF.

FAO (1983). *Food loss prevention in perishable crops (2nd edn.)*, Rome: FAO. Retrieved from http://www.fao.org/docrep/S8620E/S8620E00.htm.

FAO (2012). *Food wastage footprint. An environmental accounting of food loss and waste. Concept Note*, Retrieved from http://www.fao.org/fileadmin/templates/nr/sustainability_pathways/docs/Food_Wastage_Concept_Note_web.pdf.

FAO (2013a). *FAO statistical yearbook 2013. World food and agriculture, Rome*. Retrieved from http://www.fao.org/docrep/018/i3107e/i3107e00.htm.

FAO (2013b). Key Findings. *Save food: Global Iiitiative on food losses and waste reduction*. Retrieved August 28, 2013 from http://www.fao.org/save-food/key-findings/en/.

Foresight (2011). *The future of food and farming: Challenges and choices for global sustainability*, London. Retrieved from http://www.bis.gov.uk/assets/foresight/docs/food-and-farming/11-546-future-of-food-and-farming-report.pdf.

Gustavsson, J. et al. (2011). *Global food losses and food waste—extent, causes and prevention*, Rome. Retrieved from http://www.fao.org/docrep/014/mb060e/mb060e00.pdf.

Hall, K. D. et al. (2009). The progressive increase of food waste in America and Its environmental impact. *PloS one*, *4*(11), 9–14. Retrieved from http://www.plosone.org/article/info%3Adoi%2F10.1371%2Fjournal.pone.0007940.

Hoornweg, D., & Bhada-Tata, P. (2012). *What a waste a global review of solid waste management*, Washington, DC. Retrieved from http://www-wds.worldbank.org/external/default/WDSContentServer/WDSP/IB/2012/07/25/000333037_20120725004131/Rendered/PDF/681350WP0REVIS0at0a0Waste20120Final.pdf.

IEEP et al. (2010). *Final report—supporting the thematic strategy on waste prevention and recycling*, Retrieved from http://ec.europa.eu/environment/waste/pdf/FinalReportfinal25Oct.pdf.

ISWA (2009). *Waste and climate change ISWA white paper*, Vienna. Retrieved from http://www.iswa.org/fileadmin/user_upload/_temp_/Small_GHG_white_paper_01.pdf.

Kemp, R., & van Lente, H. (2011). The dual challenge of sustainability transitions. *Environmental Iinovation and societal transitions*, *1*(1), 121–124. Retrieved August 22, 2013 from http://linkinghub.elsevier.com/retrieve/pii/S2210422411000128.

Lipinski, B., et al. (2013). *Reducing food loss and waste*, Washington, DC. Retrieved from http://www.worldresourcesreport.org.

Neubauer, A. (2007). *Convergence with EU waste policies short guide for ENP partners (pp. 1–34)*. Retrieved from http://ec.europa.eu/environment/enlarg/pdf/pubs/waste_en.pdf.

NSW Office of Environment and Heritage Australia (2011). *Food waste avoidance benchmark study*, Sydney South NSW. Retrieved from http://www.lovefoodhatewaste.nsw.gov.au/portals/0/docs/11339FWABenchmarkstudy.pdf.

Parfitt, J., Barthel, M. & Macnaughton, S. (2010). Food waste within food supply chains: Quantification and potential for change to 2050. *Philosophical transactions of the royal society of London. Series B, Biological sciences*, *365*(1554), 3065–3081. Retrieved February 28, 2013 from http://www.pubmedcentral.nih.gov/articlerender.fcgi?artid=2935112&tool=pmcentrez&rendertype=abstract.

Raven, R. (2007). Co-evolution of waste and electricity regimes: Multi-regime dynamics in the Netherlands (1969–2003). *Energy Policy*, *35*(4), 2197–2208. Retrieved August 21, 2013 from http://linkinghub.elsevier.com/retrieve/pii/S0301421506002849.

Reisinger, H. et al. (2011). *Evolution of (bio-) waste generation/prevention and (bio-) waste prevention indicators*, Paris. Retrieved from http://ec.europa.eu/environment/waste/prevention/pdf/SR1008_FinalReport.pdf.

U.S. EPA (2009). *Opportunities to reduce greenhouse gas emissions through materials and land management practices*, Retrieved from http://www.epa.gov/oswer/docs/ghg_land_and_materials_management.pdf.

U.S. EPA (2012a). Putting surplus food to good use (pp. 1–2). Retrieved from http://www.epa.gov/wastes/conserve/pubs/food-guide.pdf.

U.S. EPA (2012b). Solid waste management hierarchy. *Wastes—non-hazardous waste—municipal solid waste*. Retrieved August 26, 2013 from http://www.epa.gov/wastes/nonhaz/municipal/hierarchy.htm.

U.S. EPA (2013a). Food recovery challenge. *Wastes—resource conservation*. Retrieved August 7, 2013 from http://www.epa.gov/epawaste/conserve/smm/foodrecovery/index.htm.

U.S. EPA (2013b). History of RCRA. *Wastes—laws and regulations*. Retrieved August 26, 2013 from http://www.epa.gov/epawaste/laws-regs/rcrahistory.htm.

UK Department of Energy and Climate Change & Defra (2011). *Anaerobic digestion strategy and action plan*, London, UK. Retrieved from https://www.gov.uk/government/uploads/system/uploads/attachment_data/file/69400/anaerobic-digestion-strat-action-plan.pdf.

UNEP (2010). *Waste and climate change: Global trends and strategy framework*, Osaka/Shiga. Retrieved from http://www.unep.or.jp/ietc/Publications/spc/Waste&ClimateChange/Waste&ClimateChange.pdf.

UNEP (2011). Waste: Investing in energy and resource efficiency. In *Towards a green economy: Pathways to sustainable development and poverty eradication (pp. 286–329)*. Retrieved from http://www.unep.org/greeneconomy/Portals/88/documents/ger/ger_final_dec_2011/8.0-WAS-Waste.pdf.

Vito, BIO Intelligence Service & Arcadis (2011). *Assessment of feasibility of setting bio-waste recycling targets in EU, including subsidiarity aspects*, Retrieved from http://ec.europa.eu/environment/waste/compost/pdf/Biowaste_recycling_targets_final_final.pdf.

Waarts, Y. et al. (2011). *Reducing food waste: Obstacles experienced in legislation and regulations*, Wageningen. Retrieved from http://edepot.wur.nl/188798.

World Economic Forum (2009). *Driving sustainable consumption value chain waste driving sustainable consumption value chain waste (pp. 1–8)*. Retrieved from http://www.weforum.org/pdf/sustainableconsumption/DSCOverviewBriefing—ValueChainWaste.pdf.

WRAP (2008). *The food we waste*, Retrieved from http://wrap.s3.amazonaws.com/the-food-we-waste.pdf.

WRAP (2012). *Decoupling of waste and economic indicators*, Retrieved from http://www.wrap.org.uk/sites/files/wrap/DecouplingofWasteandEconomicIndicators.pdf.

Chapter 3
Causes of Food Waste Generation

Experts name a variety of causes of food waste generation. Most of them are caused by certain behaviours of actors (i.e. producers, distributors, retailers, consumers), the need to follow regulations or for problems in food supply chains. In particular, problems associated with financial and technological aspects and/or related to the existing waste policies and legislation, have a major influence on the extent to which food waste is produced. It would be wrong to say that food waste is a product of the 21st century. In the 1940s researchers listed the following causes of food waste: spoilage as a result of improper food handling, later, packaging, and transportation, overstocking, plate scraps and increased portion sizes (Kantor et al. 1997; Griffin et al. 2009). Today, the following causes are the subject to experts' main concern and require immediate actions. Due to their relevance, each individual cause is addressed on this work.

3.1 Consumer Behaviour

Consumer behaviour is seen as one of the main causes of food waste in medium- and high-income countries (FAO 2012). Consumers are thought to be the largest contributors to waste along the food chain (Griffin et al. 2009; Parfitt et al. 2010; Gustavsson et al. 2011; Value Chain Management Centre 2012). According to the results of the research in the UK, food waste constitutes 60 % of food discarded overall (WRAP 2009; BIO Intelligence Service 2010). Over a quarter of this waste is still in its original packaging (Sonigo et al. 2012). Another study conducted in the EU has also indicated that food waste constitutes a significant proportion of the household and municipal waste streams (Arcadis 2010).

In 1997 Kantor et al. pointed out the following reasons for household food waste in the USA: overpreparation, preparation discard, plate waste, spoiled leftovers, breakage, spillage, and package failure, either in the home or en route from the point of purchase (Kantor et al. 1997). Earlier, in 1987, a study conducted by a university in the USA suggested that a lack of consumer education regarding food safety, insufficient knowledge for interpretation of package dating information,

W. Leal Filho and M. Kovaleva, *Food Waste and Sustainable Food Waste Management in the Baltic Sea Region*, Environmental Science and Engineering, DOI 10.1007/978-3-319-10906-0_3

expiration codes as well as confusion about an influence of quality defects on edibility, are the reasons for discarding food in households (Kantor et al. 1997). Later, excess purchases as a result of poor purchase planning, failure to use food before expiry dates, improper storage, not eating leftovers, declining knowledge of how to use leftovers and letting edible food go off either untouched or in opened packets (UK Government 2010; Foresight 2011; Value Chain Management Centre 2012; FAO 2012) were added to the list.

In 2008, the results of the WRAP study pointed out reasons for food thrown away that could have been eaten if it had been managed better (Fig. 3.1). The most common are:

- Food that has been prepared and served but not eaten—accounts for more than 1.2 million tonnes of food waste annually, mainly attributable to pre-prepared food (48 %).
- Food which has past its date—accounts for more than 800 thousand tonnes of food waste annually, mainly attributable to bread (15 %) and salads (14 %).
- Food that looked bad—accounts for nearly 470 thousand tonnes of food waste annually and mainly attributable to fresh fruit (29 %) and bread (28 %) (WRAP 2008).

The study conducted in Australia entitled 'leaving food in the fridge and/or freezer too long,' and 'not finishing of meals by members of a household,' as the most frequent causes of food waste as well as 'the length of time for safely storing different types of food,' 'lack of time,' and 'organisation and appropriate storage containers' (NSW Office of Environment and Heritage Australia 2011). The table below (Table 3.1) summarises the main findings of the study.

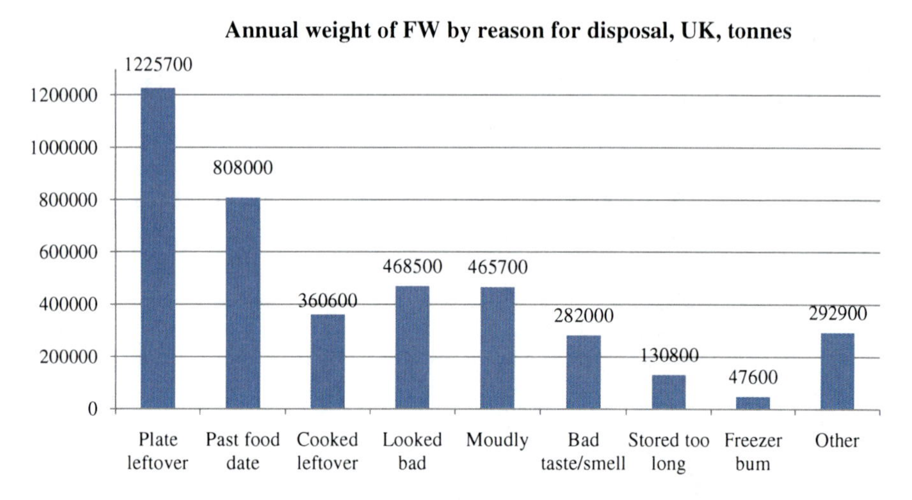

Fig. 3.1 The estimated annual weight (tonnes) of food waste by a reason for disposal (WRAP 2008)

Table 3.1 Causes of household food waste in Australia (NSW Office of Environment and Heritage Australia 2011)

Main causes of household food waste	Causes of			
	Buying too much food	Cooking too much food	Not storing food correctly	Not using leftovers
Food is left too long in the fridge and freezer	We think we need more than we actually do	Preferable to serve too much rather than not enough	I am unsure about the best way to store different food types	Forget about leftovers in the fridge and/or freezer
Some household members do not always finish their meal	We are tempted by supermarket specials e.g. 2 for 1 deals	Find it difficult to estimate how much to cook per person	Tend to leave food products in the original packaging	Do not like eating leftovers
Food goes off before the 'use by' or 'best before' date	We do not check the cupboard or fridge before shopping	Find it difficult to know how to cook the right portion sizes	Lack of time and organisation	I am unsure about how to use leftover individual/ assorted ingredients
Food bought on sale does not always last long enough	Size of food portions and packages is too large	One or more household members have different food preferences or special dietary needs	Food goes off before the 'use buy' or 'best before' date	Health concerns about eating leftovers
We cook too much food	We like to have more food or ingredients available than not enough	Not sure how many people will be home for meals	Do not have appropriate storage containers	
Family members change their plans (e.g. they do not turn up for dinner)	We do not write a list	Lack of time or organisation to plan ahead e.g. no meal plan	Do not read storage instructions	
We do not tend to use leftover ingredients in other meals	We forget to take our list	I am unsure about what visitor's food preference will be		
We do not check the fridge, freezer and cupboard before going shopping	Lack of time or organisation to plan ahead e.g. no list, no meal			

(continued)

Table 3.1 (continued)

Main causes of household food waste	Causes of			
	Buying too much food	Cooking too much food	Not storing food correctly	Not using leftovers
We buy too much food	We like fresh ingredients and do not keep older ingredients			
We like to eat the freshest food possible				
We tend not to plan meals in advance				
We are generally too busy to cook meals that we planned				

Consumer preferences also have a significant influence on the amount of food waste that is accumulated. Buying behaviour and purchasing decisions might be driven by promotions, special offers such as 'buy two, get one free,' a growing variety of products, and an increase in consumer choices. All these contribute to wasteful behaviour and result in an excessive amount of purchased food (Schneider 2008; Viel 2011; Lin et al. 2013). Consumers do want to have as big a choice as possible, therefore they expect a wide range of products to be available in stores at any point in time (e.g. 10 min before a store is closed) (Gustavsson et al. 2011). Furthermore, at home, people choose to cook more food that is needed because they prefer to serve to much food to serving not enough food. Cultural factors such as displaying wealth by leaving food uneaten, or only eating specific parts of food (Foresight 2011) also have an impact on food waste generation.

3.2 Lack of Awareness

Despite growing environmental awareness within the EU, until recent years the problem of food waste has not been recognised amongst prioritised issues. People do not think about what and how much food they throw away. Such wasteful behaviours can be entirely unconscious (BIO Intelligence Service 2010). For example, according to the 'Eurobarometer study' on "Attitudes of Europeans towards resource efficiency" (Eurobarometer 2011), 11 % of the EU citizens say

Table 3.2 Estimated percentage of food that goes to waste (Eurobarometer 2011)

More than 50 %	31–50 %	16–30 %	15 % or less	None	DK/NA
1	3	13	71	11	1

they throw no food away, 71 % think they throw away 15 % or less of the food that they bought, and only 1 % said that it was more than 50 % of the food that they bought (Table 3.2).

At the same time in the UK, 84 % of households feel they do not waste food (Viel 2011). The research findings in the Netherlands, Australia, USA and Brazil also have shown that consumers are insufficiently aware of the food waste they cause (Waarts et al. 2011; Parfitt and Barthel 2011; NSW Office of Environment and Heritage Australia 2011; Hodges et al. 2011). Such wasteful behaviour is regarded as being 'natural' within a 'culture of abundance' (Foresight 2010). Moreover, consumers do not appear to be concerned about food waste (Hodges et al. 2011). For example, a study in Australia showed that the level of concern about wasting food was lower than concern over wasted electricity or interest paid on credit cards (NSW Office of Environment and Heritage Australia 2011).

However, the problem does not only reside in the lack of awareness of the amount of food waste generated individually, but also in the fact that consumers are not aware of environmental (Waarts et al. 2011) and economic (Hodges et al. 2011) consequences of the food waste problem. This means, amongst other things, that they do not know about the financial benefits of purchasing food more efficiently (BIO Intelligence Service 2010). Besides, consumers in general do not feel personal responsibility for food waste (WRAP 2009; Lin et al. 2013).

Companies also try to avoid the issue of food waste. One way is to 'call' food waste something different, for example, by-products, losses, etc. Another way is to not regard food that has been sent for recycling, composting or donated to charity organisations as 'waste'. Such camouflaged activities alter companies' waste reports, 'help' to ignore their responsibility for food waste accumulation and avoid 'unnecessary' investments.

3.3 Labelling

The function of food product labelling is to ensure consumer safety and inform their decision making (BIO Intelligence Service 2010). However, at the same time, it plays a negative role by encouraging food waste (Institution of Mechanical Engineers 2013). One of the main problems associated with food labels is consumers' misunderstanding and mis- or poor interpretation of the meaning of the date labels such as 'best before', 'use-by', 'sell-by' (Sonigo et al. 2012). Generally, consumers do not see a difference between the 'best before' and 'use-by' dates on packaging. A piece of research undertaken in the UK showed that 45–49 % of consumers misunderstand the meaning of these two terms and this leads to 1 million tonnes of food waste (BIO Intelligence Service 2010).

The 'sell-by' or 'display until' date labels tell a store how long to display a product. The 'use-by' date label means the last date recommended for the use of a product from a food safety perspective (Lipinski et al. 2013). Whereas 'best before' indicates the period during which food is of optimum quality and does not pertain to the safety of a product (Foresight 2011; Lipinski et al. 2013). Another part of the confusion surrounding product dating is that there are a number of different terms that might appear on packages (Lipinski et al. 2013). In addition, there is a tendency among consumers to leave a safety margin and discard food even before date on a label, because they believe that date labels indicate a product's safety (BIO Intelligence Service 2010; Value Chain Management Centre 2012). Consumers also do not clearly understand the significance of storage instructions which results in unnecessary food waste (UK Government 2010).

The next aspect raising concern is a lack of consistency in using date labels terms in the Member States (BIO Intelligence Service 2010). The research findings in the Netherlands revealed the problem of an application of different expiration dates for the same type of product as well as a lack of clarity about what is permitted after the expiration date has passed. Such situations enforce businesses to remove food from thir shelves when it is not necessary, according to a product liability (Waarts et al. 2011).

Business also generates food waste because of incorrect labelling or/and a lack of clarity in interpretation of date labels terms. In this case, if the information on a label is not correct (or is presumed to be incorrect), a product may not be sold and is taken off the market, despite the fact that it is safe for consumption. Repacking or relabeling costs of such products are much higher than costs of their disposal (Waarts et al. 2011).

Additionally, in order to limit their own risk related to product liability and avoid legal actions producers and retailers leave safety margins by setting conservative 'best before' dates for products and do not provide any guarantee after opening (Waarts et al. 2011; Institution of Mechanical Engineers 2013).

Moreover, many retail companies are confused about what is and what is not legally permitted after the 'best before' date has passed. This also leads to food products being taken off the shelves unnecessarily (Waarts et al. 2011).

All points mentioned above indicate that current food product labelling is not functioning optimally and makes date labelling one of the subjects in food waste prevention programs (BIO Intelligence Service 2010).

3.4 Aesthetic Standards

Consumer preferences and insufficient or a lack of knowledge are not the only reasons for food to be thrown. Some reasons, particularly in developed countries, are a result of consumers' high expectations. One of such causes is product aesthetic standards. Today, food that is perfectly safe and fit for consumption might be rejected only because of physical or aesthetic defect, such as being the wrong shape or size; being broken or having a cleft or a blemish (Gustavsson et al. 2011). The obsession about cosmetic quality standards does not have any relation to the wholesomeness of food (Parfitt and Barthel 2011). To meet consumer expectations supermarkets set up strict quality standards on weight, size, shape and appearance. As a result, substantial quantities of perfectly edible fruits and vegetables are rejected by supermarkets already at the farm gate due to these artificial standards (Gustavsson et al. 2011). For example, up to 30 % of the UK's vegetable crop is never harvested as a result of such practices. Globally, in this way, retailers generate 1.6 million tonnes of food waste annually (Institution of Mechanical Engineers 2013).

The food processing industry follows the same trend as well. The term 'culling' is applied to the process of the removal of products based on quality or appearance criteria, including specifications for size, colour, weight, blemish level etc. (Value Chain Management Centre 2012). Furthermore, food processing lines often carry out trimming to ensure the end product is the right shape and size. Such trimmings are usually disposed of (Gustavsson et al. 2011). But even before this stage, producers selectively leave small, misshapen, or otherwise blemished produce in the field during the harvesting stage, since these commodities would likely be discarded in the packing shed or processing plant (Kantor et al. 1997).

A technological aspect also 'contributes' to food waste generation. Errors during food processing that end with final products with the wrong weight, shape or appearance, or damaged packaging, without affecting the safety, taste or nutritional value of the food, lead to such products to be discarded already at the production stage due to the aforementioned aesthetic standards (Gustavsson et al. 2011).

3.5 Food Merchandising

The main goal of any marketing strategy is to increase the selling rates. Promotion activities regarding food products are not an exception. Therefore marketing as a tool also plays a significant role in encouraging wasteful consumer behaviour (Value Chain Management Centre 2012). Offers such as, 'buy one get one free' coupons, super-sized portions, as well as bulk discounts and high-pressure advertising campaigns encourage consumers to buy excessive quantities regardless of their needs, which leads to substantial food waste in the home (BIO Intelligence Service 2010; Gustavsson et al. 2011; Hodges et al. 2011; Value Chain Management Centre 2012; Institution of Mechanical Engineers 2013). In some way these types of marketing activities can be considered not only as an instrument to generate higher profit but also as a way to avoid possible food waste by transferring it to a consumer side, making consumers responsible for the extra food waste generated.

The issue of correct portion sizing raises a lot of concern as well. Today, despite a high variety of products it is almost impossible to find a product with the required packaging size (e.g. a half kilo packaging instead of one or two kilo packaging). Usually it happens because companies rarely take into account the different dietary and energy requirements of their customers (e.g. men, women, children) (Foresight 2010). A similar problem can be also found in the food service and hospitality industries. By offering only single options of portion size or buffet, restaurants, canteens and etc. 'enforce/encourage' customers to leave leftovers on their plates because those are unable to finish a meal (Lipinski et al. 2013).

3.6 Legislation/Regulations as an Obstacle

Legislation appears to be not only one of the tools for a solution to the problem of food waste, but also one of the obstacles for food waste reduction. Mostly, legislative or regulatory barriers occur during the stage of treatment of accumulated food waste. There are cases when consumers and/or distributors would like to take some measures to prevent food waste but it was unsuccessful because of regulations, mainly regarding food safety. For example, in France a school could not redistribute the bread left behind at lunch because a town forbid it (Viel 2011). Stores also prohibit the use of food intended to be discarded because, according to one of the articles of the French Environmental Code, the store owner will be held responsible in a court case if food waste causes poisoning on a person who retrieves it (Viel 2011). On the European level such situations are regulated by the Product Liability Directive (Council Directive 85/374/EEC 1985). According to the document, a producer/importer/supplier may be liable in a court case when a defective product causes damage to a consumer. Therefore, a discard (i.e. waste) option is considered by companies as the safest.

In the Netherlands, companies perceive the stricter hygiene codes than the legislation they are based on as a legal obstacle which results in unnecessary food waste. The hygiene rules contain very large safety margins and the issue of food safety is sometimes taken too far by the actors (Waarts et al. 2011). There are also time limits such as 'the 2-h guarantee' on unrefrigerated products offered for sale. The regulation means that products which normally need to be stored refrigerated, may be offered for sale for a maximum of 2 h and must afterwards be thrown away, whether packaged or unpackaged (Waarts et al. 2011). The European regulations describe general and specific marketing standards for different types of food products. There are specific marketing standards with rules for: quality (minimum requirements, classification into classes), grading (size), tolerance, packaging and appellation.

3.6.1 European Marketing Standards

The European Marketing Standards regarding fresh fruit and vegetables (FFV) are perceived as one of the causes of food waste in the food chain (Waarts et al. 2011). Marketing standards are requirements that intend to guarantee EU consumers a particular minimum quality of the products they buy. They are mainly related to quality and labelling of products at the retail stage (CBI Netherlands 2012). The regulation sets the general and specific marketing standards for FFV. It has been evolved since 2007. In 2009, the number of fresh products covered by the European marketing standards was reduced from 36 to 10. Today, the specific standards are applied to the following 10 products: apples, citrus fruit, kiwifruit, lettuces, curled leaved and broad-leaved endives, peaches and nectarines, pears, strawberries, sweet peppers, table grapes, tomatoes (Commission implementing Regulation (EU) No 543/2011 (2011)). In addition, regulations dictated the exact shape, size and appearance for fruits and vegetables were abolished in the amending version. However, the current document still defines such product-specific requirements as classification ('Extra' class, Class I, Class II), minimum size and quality, sorting, minimum maturity requirements for each of these 10 products. Products which deviate from the marketing standard might not be traded as fresh products within the chain (Waarts et al. 2011).

UNECE has also developed standards concerning the marketing and commercial quality control of FFV, including the 10 types covered by the specific EU marketing standards. The specific EU marketing standards for individual products must be in line with the relevant UNECE standards. These standards are updated once a year (European Commission 2012). The UNECE standards are stricter than the EU standards and are therefore preferred by many EU buyers (CBI Netherlands 2012). The elimination of a number of marketing standards has had no big effect (Waarts et al. 2011). It is still difficult to find odd shaped fruit and vegetables in EU supermarkets. The reasons are usage of old standards as private quality requirements, slow adaptation of the current supply chains to new possibilities (Waarts et al. 2011), EU consumers are still sensitive for a certain shape and appearance of FFV (CBI Netherlands 2012).

3.7 Companies Private Standards and Reputation

Today, the old original European marketing standards are adopted by some chain actors in the fruit and vegetables trade in the form of a private classification system (Waarts et al. 2011). One of the main reasons for such behaviour is a company's reputation, which has a direct impact on its profit. Therefore companies make great efforts to protect and keep it up. In the food industry, in order to avoid any legal actions and at the same time to insure their reputation, companies set up private standards which are stricter than the standards set by legislation. For example, in the

Netherlands suppliers prefer to take the following precautionary measures in order to avoid a wrong delivery of food (e.g. in spoiled condition), which might affect their image: (Waarts et al. 2011)

- Place a relatively early expiration date on the products (the true shelf life is often longer);
- Take precautionary measures to deliver a standard product quality to the processors;
- Adopt safety margins for ordered lunches (caterers)

Another reason for private standards is the technology, which depends on the quality standards. Usually standard processing lines fit for products with same size and shape. The studies in the Netherlands showed that, for example, a fruit grower prefers to use as many as possible 'A class' (first class) products, in order to avoid losses in an operation, because processing of deviant forms or remainders is not efficient, since the machines are not set up for them (Waarts et al. 2011). The same is applied to the logistic processes. Out-of-standard shape products, mostly vegetables are not offered for sale in the regular supermarkets, because in the logistic process, it is much more efficient to pack products with the same size and shape than with deviant forms (Waarts et al. 2011).

Other types of private standard relate to expiration dates, lower Maximum Residue Levels (MRLs), upper legal levels of a concentration for pesticide residues in food (European Food Safety Authority 2013) and delivery temperature. These standards are also stricter than legally required. For example, for fresh products, companies often demand a particular expiration term. If the 'best before' date is too short, the product will not be bought (Waarts et al. 2011). Another example is when supermarkets use 50 % of the legal MRLs as their standard because measurement results can be out by 50 %. As a result, fewer crop protection agents are used and more waste is caused by putrefaction (Waarts et al. 2011). Moreover, in answer to increasing consumer concerns about food safety there exists a consortia of EU retailers and supermarkets that together developed additional standards for FFV such as GlobalGAP to assure a minimum safety guarantee (CBI Netherlands 2012). The GlobalGAP certification is a requirement of the largest retail companies and wholesalers in the EU. Furthermore, there are supermarkets such as Tesco in the UK and major French supermarkets that have developed their own company-specific standards or 'supplier criteria' (CBI Netherlands 2012).

3.8 Overproduction and Excess Stock

Another cause of food waste generation which has also a relation to a company's image is overproduction or overstocking. By following incorrect estimates of sales and financial stimulus to achieve the highest possible turnover, supermarkets order, restaurants, canteen or caterer prepare extra quantities in order to avoid the situation of 'sorry, sold out' (Waarts et al. 2011).

In the Netherlands primary producers see overproduction of fruits and vegetables as the most important cause of food waste (Waarts et al. 2011). Overproduction has a number of reasons such as a lack of agreements about the quantities to be produced, which results in a significant difference between supply and demand. It also has an economic consequence. Bigger quantity of a product supplied to a market causes falling prices, therefore in order to prevent such situations and keep the price artificially high fruit and vegetables are sometimes destroyed (Waarts et al. 2011).

In the food manufacturing sector there is deliberate over-production in response to the uncertainties imposed by retailers. If a retailer changes an order at the last minute, the supplier has to cover the increase/decrease in goods demanded or face the risk of being de-listed (Foresight 2010). Bakeries produce more than necessary just to create impression of product availability.

Poor inventory management reflected in inaccurate ordering or forecasting of demand, the 'take-back' system and last minute order cancellation, that entitle retailers due to supply contracts to return stock to their suppliers once it has reached a specified amount of residual shelf-life remaining (e.g. 75 %) lead to unnecessary excess stock generated by one of the food chain actors (Defra 2007). As a result suppliers are limited with options on how to dispose the returned stock. This causes significant volumes of food waste to be discarded instead of to be sold with no food safety concerns (BIO Intelligence Service 2010). Published studies on supermarket discard show that overstocking or/and improper stock rotation of seasonal items like Easter eggs are also reasons that retailers discard food (Kantor et al. 1997).

Results of the research in several EU countries show that during the planning stage producers/importers are unable to foresee the quantity they can sell. They order/produce too much—and this becomes food waste. Supermarket chains give out sale forecasts, however, they do not take any responsibility for the provided data. Suppliers order/produce according to these forecasts but supermarkets might buy only a small part of forecasted quantity—the rest is food waste (Ujhelyi 2013).

3.9 Food Prices/Financial Incentives

Despite a worldwide concern about rising food prices, which doubled from 2006 to 2008 (Beddington et al. 2012) and statements about increasing pressure on the most vulnerable households who spend a considerably greater proportion of their income on food (European Commission Communication COM (2009) 591), the most incredible point is that the current prices are still too low and therefore large amounts of food are thrown away. However, rising prices also indicate an excess of demand over supply i.e. insufficient amounts of food. Over the years the proportion of disposable income spent on food by households has declined (Escaler and Teng 2011). Today a policy of 'cheap food' is an attribute of the most developed countries. For example, average family in the USA spend only 6.6 % of total consumer expenditures on food, in Norway this number is 13.2 % (Fig. 3.2),

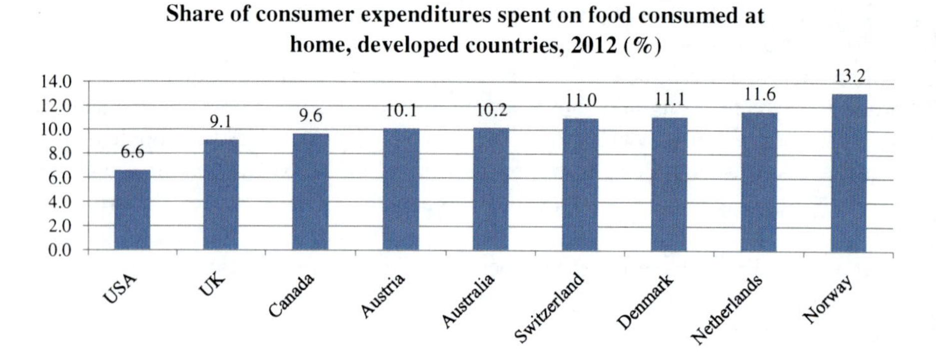

Fig. 3.2 Share of consumer expenditures spent on food consumed at home in developed countries (%) in 2012 (ERS/USDA 2013)

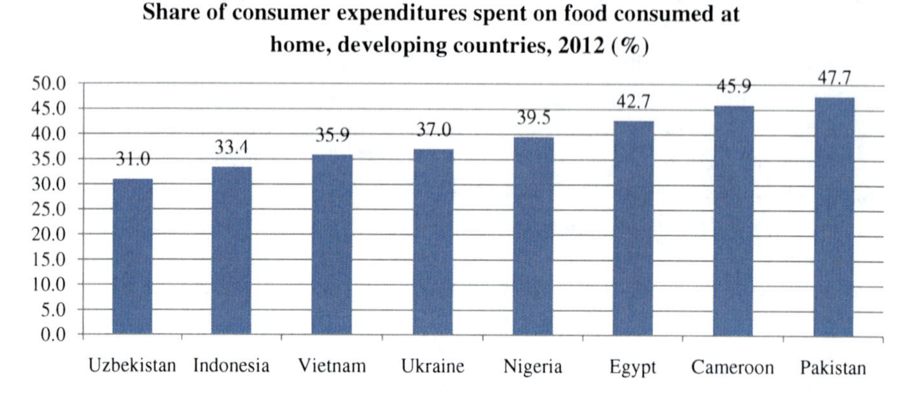

Fig. 3.3 Share of consumer expenditures spent on food consumed at home in developing countries (%) in 2012 (ERS/USDA 2013)

whereas the share of expenditures for food in developing countries such as Vietnam and Cameroon reach 35.9 and 45.9 % respectively (Fig. 3.3).

Such a gap in expenditures indicates that in developed countries consumers undervalue food resources (BIO Intelligence Service 2010) and, generally, food regarded as the least costly resource (Institution of Mechanical Engineers 2013).

In the 1970s, several studies pointed out that the availability of cheap food encourages overbuying and hoarding behaviours that result in waste (Griffin et al. 2009). Consumers do not feel the impact of waste in their wallets. Hence they have weak financial incentives to minimize food waste (Waarts et al. 2011; Hodges et al. 2011).

The negative correlation between the price of food and the amount of food waste generated has also been supported by results of the study conducted in 2011 by Coldiretti-SWG in Italy. The study showed that Italians have reduced food waste by

57 % because of the economic crisis. Today three out of four Italians manage their spending more carefully than they did before (Ujhelyi 2013).

A lack of financial incentives demotivates companies to reduce their current amounts of food waste. The same is true about residual flows. If a residual flow yields no income and/or waste disposal is cheap, companies will not look for other ways of reusing their residual flows (Waarts et al. 2011). The processing industry also lists a lack of demand in the market, the absence of finance for improved value utilisation of the waste flow and the unwillingness to invest in separating waste flows because the investment would not pay for itself (Waarts et al. 2011) as obstacles to reuse waste flows. Currently, there are also no financial incentives for supermarket retailers to reduce waste levels. One of the reasons for thisis low cost of disposing of excess supply in comparison to prices of other alternatives (Parfitt and Barthel 2011).

Price is also very important factor. There are situations when crops might go unharvested, only because the price is too low and does not cover the labour and transport costs required to pick and sell that crop. As a result it becomes economically rational for a farmer to let that food be wasted (Lipinski et al. 2013). The caterers are also not always impacted by food waste in financial terms. In 40 % of catering contracts, waste disposal is paid for. Thus waste remains the property of customer and not of the caterers (Waarts et al. 2011). It is also important to note that the European Parliament take steps in lowering of food prices through a reduction in food waste. On the one hand it will improve the access to food by poorer segments of the population (European Parliament Resolution (2011/2175 (INI)) 2012), however, on the other hand it might increase amount of food waste generated as a result of the rebound effect.

3.10 Technical Factors

There are a number of technical factors effecting food waste generation, which arise during storage, transportation, packaging and distribution processes. Imperfection, availability or unavailability of technology also has an impact on food waste accumulation. For example, waste during the harvesting stage might occur because of an inability of current technology to discriminate between immature and ripe products or of mechanized harvesters retrieve the entire item (Kantor et al. 1997). Another example could be an inaccessibility of existing technologies due to various reasons, most of which are economical. Companies are not ready to invest in the latest available technologies which would reduce food waste because it will be not cost-efficient in countries of their residence.

3.10.1 Storage

Inappropriate storage conditions leads to food waste throughout the supply chain and in the household (BIO Intelligence Service 2010). At the consumer level there is a lack of identification of retention periods (Viel 2011) as well as insufficient knowledge about methods of food storage, a lack of consumer attention to storage labels and the absence of storage guidance (BIO Intelligence Service 2010). WRAP reports that over two million tonnes of food is not being stored correctly in the UK (BIO Intelligence Service 2010). Technical problems such as interruption of the cool chain or incorrect storage temperature because of a lack of knowledge of ideal storage conditions also lead to food waste (Waarts et al. 2011).

3.10.2 Stock Transportation

Stock transportation is inherently linked to storage conditions. Extreme changes in temperature during transportation can spoil or shorten the shelf life of food products, particularly sensitive to temperature conditions (BIO Intelligence Service 2010). There are also logistical limitations associated with 'imperfect' size and shape of a product. The logistical process is usually perfected, therefore to keep it efficient only products of same size and shape are packed (Waarts et al. 2011). However, it is important to note that transport and distribution accounts for 'only' approximately 3 % of the food wasted along the supply chain (Escaler and Teng 2011; Value Chain Management Centre 2012).

3.10.3 Poor Packaging

Poor packaging performance resulting in damage to food products will lead to product product being discarded. Damage to the product's primary or secondary packaging also means that the product will often be discarded, while the food itself is unharmed (BIO Intelligence Service 2010). Sometimes due to disruptions and production errors, packaging is not uniform in its contents, as a result such packaging and its content is rejected and often destroyed (Waarts et al. 2011). In addition, in the case of wrong packaging there is technical challenge with regard to the unpacking of packaged products, because it is time consuming, requires a lot of effort and is very costly (Waarts et al. 2011).

Thus, each of the numerous causes that lead to food waste generation is not only caused by actors of the food chain, but also by other causes of food waste. The situation is very similar to the domino effect. For example, consumer expectations of the availability of a product induces the following sequence of activities resulting in food waste: false sales projections, overproduction, oversupply and, finally, food disposal. Table 3.3 summarises the main causes of food waste and their implications. This large variety of aspects and affected areas indicates on a necessity of involvement of a big number of actors from all stages of the food supply chain to tackle the problem.

Table 3.3 Main causes of food waste generation

Problem	Implications
Consumer behaviour	• Excess purchases, as a result of poor purchase planning
	• Failure to use food before expiry dates
	• Improper storage
	• Leaving food in the fridge and/or freezer too long
	• Food has been prepared and served but not eaten
	• Declining knowledge of how to use leftovers
	• Letting edible food go off either untouched or in opened packets
	• Lack of consumer education regarding food safety
	• Insufficient knowledge for interpretation of food date labels
	• Confusion about an influence of quality defects on edibility
	• Purchasing decisions, driven by promotions, special offers: 'buy two, get one free', a growing variety of products
	• Expectations of a wide range of products to be available in stores at any point of time
	• Cultural factors: displaying wealth by leaving food uneaten
Lack of awareness	• People do not think about what and how much food they discard
	• Consumers are not concerned about food waste
	• Consumers are not aware of negative environmental and economic consequences of food waste
	• Consumers do not feel personal responsibility in relation to food waste
	• Companies 'call' food waste differently, e.g. by-products or losses
	• Food, sent for recycling, composting or donated, is not regarded as 'waste'
Labelling	• Misunderstanding and mis- or poor interpretation of the date labels: 'best before', 'use-by', 'sell-by'
	• Consumers leave a safety margin and discard food even before date on a label
	• Consumers do not clearly understand the significance of storage instructions
	• A lack of consistency in using date labels terms in the EU countries
	• A product is taken off the market as a result of incorrect (or presumed to be incorrect) information on a label
	• Producers and retailers set conservative 'best before' dates to limit risk related to a product liability and avoid legal actions
Aesthetic standards	• Food is rejected because of wrong shape or size; being broken, having a cleft or a blemish
	• Supermarkets set up strict quality standards on weight, size, shape and appearance

(continued)

Table 3.3 (continued)

Problem	Implications
	• Perfectly edible FFV are rejected by supermarkets already at farms
	• Food processing lines carry out trimming to ensure the end produc is in the right shape and size. Trimmings are usually disposed of
	• The harvesting stage: producers selectively leave small, misshapen, or otherwise blemished produce in the field
	• Final products are discarded because of wrong weight, shape or appearance, or damaged packaging, as a result of errors during the food production stage
Food merchandising	• Coupons, 'buy one get one free', super -sized portions, bulk discounts, high-pressure advertising campaigns encourage a purchase of excessive quantities
	• Unavailability of products with needed packaging size
	• Single option for portion size or buffet offered by restaurants and canteens, resulted in leftovers on consumers' plates
Legislation/regulations as an obstacle:	• Food safety legislative or regulatory barriers during food waste treatment
	• Product Liability Directive makes a producer/importer/ supplier liable in a case, when a defective product causes damage to a consumer
	• Hygiene rules with very large safety margins
	• The '2-h guarantee' time limits on unrefrigerated products offered for sale
European marketing standards	• Specific marketing standards: classification ('Extra' class, Class I, Class II), minimum size and quality, sorting, minimum maturity requirements
	• Applied to each of 10 products: apples, citrus fruit, kiwifruit, lettuces, curled leaved and broad-leaved endives, peaches and nectarines, pears, strawberries, sweet peppers, table grapes, tomatoes
	• Deviated from the marketing standard products are not traded as fresh products within the chain
	• UNECE standards are stricter than the EU standards and preferred by many EU buyers
Companies private standards and reputation	• To protect reputation companies set up stricter than legally required, private standard related to lower MRLs, upper legal levels of a concentration for pesticide residues in food and delivery temperature
	• Suppliers place a relatively early expiration date on the products (the true shelf life is longer)
	• Standard processing lines fit for products with same size and shape

(continued)

Table 3.3 (continued)

Problem	Implications
	• Fruit growers prefer to use as many as possible 'A class' (first class) products, because the machines are not set up for processing of deviant forms or remainders
	• The logistic process: it is much more efficient to pack products with the same size and shape than with deviant forms
	• Additional standards for FFV set by the largest retail companies and wholesalers e.g. GlobalGAP certification
Overproduction and excess stock	• Poor inventory management: inaccurate ordering or fore-casting of demand, the 'take-back' system, last minute order cancellation, stock returned to suppliers once it has reached a specified amount of residual shelf-life remaining (e.g. 75 %)
	• Overproduction as a result of the lack of agreements about the quantities to be produced
	• FFV are sometimes destroyed to keep the price artificially high
	• Deliberate overproduction in the food manufacturing sector in response to the uncertainties imposed by retailers
	• Bakeries produce more than necessary to create impression of products availability
	• Overstocking or/and improper stock rotation of seasonal items
	• Supermarkets do not take any responsibility for the provided to suppliers data
Food prices/financial incentives	• A policy of 'cheap food' is an attribute of the most developed countries
	• In developed countries consumers undervalue food resources and food regarded as the least costly resource
	• Cheap food encourages overbuying and hoarding behaviours
	• Consumers do not feel the impact of waste in their wallets
	• Lack of financial incentives for companies to reduce their amounts of food waste
	• A residual flow yields no income
	• Waste disposal is cheap
	• Absence of finance for improved value of the waste flow
	• Unwillingness to invest in separating waste flows, because the investment would not pay for itself
	• Crops go unharvested, because low price does not cover the labour and transport costs required to pick and sell that crop
	• In 40 % of catering contracts, waste disposal is paid for
	• Measures taken by the European Parliament to lower food prices through a reduction of food waste might increase its amount as a result of the rebound effect
Technical factors:	• Food waste is a result of imperfection, availability or unavailability of technology

(continued)

Table 3.3 (continued)

Problem	Implications
	• Unwillingness of companies to invest in the latest available technologies to reduce food waste because of cost-inefficiency
	• The harvesting stage: limitation of current technology to discriminate between immature and ripe products, or of mechanized harvesters retrieve the entire item
Storage	• Inappropriate storage conditions
	• Interruption of the cool chain or incorrect storage temperature because of a lack of knowledge of ideal storage conditions
	• The consumer level: lack of identification of retention periods, insufficient knowledge about methods of food storage, lack of consumer attention to storage labels and the absence of storage guidance
Stock transportation	• Extreme changes in temperature during transportation can spoil or shorten the shelf life of food products, particularly sensitive to temperature conditions
	• Logistical limitations associated with 'imperfect' size and shape of a product
	• Logistical process is efficient only for packed products of same size and shape
Poor packaging	• Product is discarded because of its damaged primary or secondary packaging
	• Packaging is not uniform in its contents due to disruptions and productions errors
	• Wrong packaging: technical challenge to unpack packaged products, because it is time consuming, requires a lot of effort and very costly

References

Arcadis (2010). *Assessment of the options to improve the management of bio-waste in the European Union—final report.* Retrieved from http://ec.europa.eu/environment/waste/compost/pdf/ia_biowaste-finalreport.pdf.

Beddington, J. et al. (2012). *Achieving food security in the face of climate change: Final report from the commission on sustainable agriculture and climate change. cGIAR research program on climate change, agriculture and food security (CCAFS).*, Copenhagen. Retrieved from http://ccafs.cgiar.org/sites/default/files/assets/docs/climate_food_commission-final-mar2012.pdf.

BIO Intelligence Service (2010). *Preparatory study on food waste across EU 27.* Retrieved from http://ec.europa.eu/environment/eussd/pdf/bio_foodwaste_report.pdf.

CBI Netherlands (2012). EU legislation: Marketing standards for fresh fruit and vegetables. (pp. 1–6). Retrieved from http://www.cbi.eu/system/files/marketintel/2012_EU_legislation_Marketing_standards_for_fresh_fruit_and_vegetables.pdf.

Commission implementing Regulation (EU) No 543/2011 (2011). *Commission implementing Regulation (EU) No 543/2011 of 7 June 2011 laying down detailed rules for the application of council regulation (EC) No 1234/2007 in respect of the fruit and vegetables and processed fruit and vegetables sectors*, Official Journal L157, 15/6/2011 (pp. 1–163). Retrieved from http://eur-lex.europa.eu/LexUriServ/LexUriServ.do?uri=OJ:L:2011:157:0001:0163:EN:PDF.

Council Directive 85/374/EEC (1985). *Council Directive 85/374/EEC of 25 July 1985 on the approximation of the laws, regulations and administrative provisions of the Member States concerning liability for defective products*, Official Journal L 210. Retrieved from http://eur-lex.europa.eu/LexUriServ/LexUriServ.do?uri=CONSLEG:1985L0374:19990604:EN:PDF.

Defra (2007). *Report of the food industry sustainability strategy champions' group on waste*, London, UK. Retrieved from http://archive.defra.gov.uk/foodfarm/policy/foodindustry/documents/report-waste-may2007.pdf.

ERS/USDA (2013). *Food expenditures*. Retrieved September 12, 2013 from http://www.ers.usda.gov/data-products/food-expenditures.aspx#.UrrPO_QW1dN.

Escaler, M. & Teng, P. (2011). "Mind the gap": Reducing waste and losses in the food supply chain food losses in the food supply chain. *The centre for non-traditional security (nts) studies, s. rajaratnam school of international studies (RSIS), NTS Insight, (6)*. Retrieved from http://www.rsis.edu.sg/nts/HTML-Newsletter/Insight/pdf/NTS_Insight_jun_1101.pdf.

Eurobarometer (2011). *Attitudes of Europeans towards resource efficiency*. Retrieved from http://ec.europa.eu/public_opinion/flash/fl_316_en.pdf.

European Commission (2012). Fruit and vegetables: Marketing standards. *Fruit and vegetables*. Retrieved October 2, 2013 from http://ec.europa.eu/agriculture/fruit-and-vegetables/marketing-standards/index_en.htm#specific-marketing-standards.

European Commission Communication COM (2009) 591, 2009. *European Commission Communication COM (2009) 591 to the European parliament, the council, the European economic and social committee and the committee of the regions on a better functioning food supply chain in Europe* (pp. 1–14). Retrieved from http://eur-lex.europa.eu/LexUriServ/LexUriServ.do?uri=COM:2009:0591:FIN:EN:PDF.

European Food Safety Authority (2013). *Maximum residue levels*. Retrieved September 30, 2013 from http://www.efsa.europa.eu/en/pesticides/mrls.htm.

European Parliament Resolution (2011/2175 (INI)) (2012). *European Parliament Resolution of 19 January 2012 on how to avoid food wastage: Strategies for a more efficient food chain in the EU (2011/2175(INI)) 2175(January)* (pp. 1–8). Retrieved from http://www.europarl.europa.eu/sides/getDoc.do?pubRef=-//EP//NONSGML+TA+P7-TA-2012-0014+0+DOC+PDF+V0//EN.

FAO (2012). *Global initiative on food losses and waste reduction*. Retrieved from http://www.fao.org/docrep/015/i2776e/i2776e00.pdf.

Foresight (2010). *How can waste reduction help to healthily and sustainably feed a future global population of nine billion people?*, London. Retrieved from http://www.bis.gov.uk/assets/foresight/docs/food-and-farming/workshops/11-608-w4-expert-forum-reduction-of-food-waste.pdf.

Foresight (2011). *The Future of Food and Farming: Challenges and choices for global sustainability*, London. Retrieved from http://www.bis.gov.uk/assets/foresight/docs/food-and-farming/11-546-future-of-food-and-farming-report.pdf.

Griffin, M., Sobal, J. & Lyson, T. A. (2009). An analysis of a community food waste stream. *Agriculture and Human Values*, *26*, 67–81. Retrieved from http://download.springer.com/static/pdf/920/art%253A10.1007%252Fs10460-008-9178-1.pdf?auth66=1364125369_6d36c30a39f962d421c6f970df9fbbf5&ext=.pdf.

Gustavsson, J. et al. (2011). *Global food losses and food waste—Extent, causes and prevention*, Rome. Retrieved from http://www.fao.org/docrep/014/mb060e/mb060e00.pdf.

Hodges, R. J., Buzby, J. C. & Bennett, B. (2011). Postharvest losses and waste in developed and less developed countries: Opportunities to improve resource use. *Journal of Agricultural Science*, *149*, 37–45. Retrieved from http://journals.cambridge.org/download.php?file=%2FAGS%2FAGS149_S1%2FS0021859610000936a.pdf&code=16a2afa2796c12d2e4a6c55bf429b91c.

Institution of Mechanical Engineers (2013). *Global food waste not, want not.*, London. Retrieved from http://www.imeche.org/Libraries/Reports/Global_Food_Report.sflb.ashx.

Kantor, L. S. et al. (1997). Estimating and addressing America's food losses. *Food Review, 1264* (202), 2–12. Retrieved from http://www.calrecycle.ca.gov/reducewaste/food/foodlosses.pdf.

Lin, C. S. K. et al. (2013). Food waste as a valuable resource for the production of chemicals, materials and fuels. Current situation and global perspective. *Energy and Environmental Science*, *6*(2), 426–464. Retrieved August 13, 2013 from http://xlink.rsc.org/?DOI=c2ee23440h.

Lipinski, B. et al. (2013). *Reducing food loss and waste*, Washington, DC. Retrieved from http://www.worldresourcesreport.org.

NSW Office of Environment and Heritage Australia (2011). *Food waste avoidance benchmark study*, Sydney South NSW. Retrieved from http://www.lovefoodhatewaste.nsw.gov.au/portals/0/docs/11339FWABenchmarkstudy.pdf.

Parfitt, J. & Barthel, M. (2011). *Global food waste reduction: Priorities for a world in transition*, London. Retrieved from http://www.bis.gov.uk/ssets/foresight/docs/food-and-farming/science/11-588-sr56-global-food-waste-reduction-priorities.pdf.

Parfitt, J., Barthel, M. & Macnaughton, S. (2010). Food waste within food supply chains: Quantification and potential for change to 2050. *Philosophical transactions of the royal society of London. Series B, Biological sciences*, *365*(1554), 3065–3081. Retrieved February 28, 2013 from http://www.pubmedcentral.nih.gov/articlerender.fcgi?artid=2935112&tool=pmcentrez&rendertype=abstract.

Schneider, F. (2008). *Wasting food – an insistent behaviour*, Alberta, Canada. Retrieved from http://www.ifr.ac.uk/waste/Reports/WastingFood-AnInsistent.pdf.

Sonigo, P. et al. (2012). *Assessment of resource efficiency in the food cycle, Final report.* Retrieved from http://ec.europa.eu/environment/eussd/pdf/foodcycle_Finalreport_Dec2012.pdf.

Ujhelyi, K. (2013). *FoRWaRD survey report.* Retrieved from http://foodrecoveryproject.eu/wp-content/uploads/2012/11/FoRWaRd-D3.3_Report_of_Analysis_of_Results.pdf.

UK Government (2010). *Food 2030.* Retrieved from http://archive.defra.gov.uk/foodfarm/food/pdf/food2030strategy.pdf.

Value Chain Management Centre, 2012. *Cut waste, grow profit.* Retrieved from http://www.valuechains.ca/usercontent/documents/CutWasteGrowProfitFINALDOCUMENTOct312pdf.

Viel, D. (2011). *Food wastage study mid-term report.* Retrieved from http://www.developpement-durable.gouv.fr/IMG/foodwastemid-termreport_VF.pdf.

Waarts, Y. et al. (2011). *Reducing food waste: Obstacles experienced in legislation and regulations*, Wageningen. Retrieved from http://edepot.wur.nl/188798.

WRAP (2008). *The food we waste.* Retrieved from http://wrap.s3.amazonaws.com/the-food-we-waste.pdf.

WRAP (2009). *Household food and drink waste in the UK*, Banbury. Retrieved from http://www.wrap.org.uk/sites/files/wrap/HouseholdfoodanddrinkwasteintheUK-report.pdf.

Chapter 4
Methods of Food Waste Reduction

The choice of methods of waste reduction should lie with the main actors of the food supply chain (Parfitt et al. 2010) and target social, cultural, economic and legal areas. The activities/projects/initiatives, focusing on changing of behaviour and attitudes towards the problem of food waste, could be realised through better education, an increase of public awareness with regard to the state of the problem, its environmental impact, methods of food waste prevention, etc. It is also important to influence a cultural perception of food i.e. to transform the idea of a low value food type as a resource, its abundance and dependence of quality on aesthetic standards. The issue of food labelling, better understanding and interpretation of the meaning of the label content require a lot of attention as well.

An application of currently available technologies and development of new ones also contribute to the solution of the problem, especially, during the harvesting, production, distribution and retail stages. However, government interventions through legislation, regulation policies and economic incentives, as well as non-governmental initiatives, are considered to be the main drivers in the implementation of measures aimed at reduction of food waste at both national and international levels.

Further discussed methods of food waste reduction, which are currently used in developed countries, could be divided into two types, in accordance with the food waste recovery hierarchy (Fig. 2.9). The first type covers methods, which still treat food waste as food, the second type includes methods that help to avoid disposal/landfill of food waste.

4.1 Public Awareness Raising/Education

Knowledge is one of the most important factors that can shape a persons behaviour. The behaviour of actors of the food supply chain is not an exception. For example, the UK government, in their new strategy for food until 2030, defined education, information and personal responsibility as one of the key points in reaching the goal to reduce food waste (UK Government 2010). Informative tools such as awareness

W. Leal Filho and M. Kovaleva, *Food Waste and Sustainable Food Waste Management in the Baltic Sea Region*, Environmental Science and Engineering, DOI 10.1007/978-3-319-10906-0_4

or/and education have an impact on every stage of the product life cycle, from design over consumer phase to waste and end-of-waste phase (Arcadis et al. 2010), this fact should be considered during development of strategies aimed at elimination or at least reduction of food waste. The analysis of the current situation talks about urgency of drawing public attention to the extent of the problem. Experts agree upon the necessity (Gustavsson et al. 2011), importance and effectiveness (BIO Intelligence Service 2010) of direct communication and awareness-raising among consumers (Parfitt et al. 2010) in order to reduce food waste. Therefore, one of the first steps that should be taken is building awareness (Foresight 2010; BIO Intelligence Service 2010; Weber et al. 2011). It could be implemented by using such instruments as awareness campaigns, training programmes, purchasing guidelines, advertising and educational initiatives (Beddington et al. 2012). The main goal is to motivate or 'enforce' actors to change their behaviour regarding food.

4.1.1 Awareness Campaigns and Informativeness

Causes of household food waste, described in the previous chapter, such as lack of awareness and knowledge on methods for avoiding food waste, date label confusion and inappropriate storage can be directly addressed through awareness and information campaigns (BIO Intelligence Service 2010). Success of the tool was proved by different programmes around Europe (Ellen MacArthur Foundation 2013). Moreover, effective communication to households on the benefits of food waste recovery is seen as essential to maximise their participation in the process (Bridgwater and Parfitt 2009; Lamb and Fountain 2010).

Governments could also organise campaigns to inform consumers and make them more aware of their wealth and the value of food (Waarts et al. 2011). One of the possible solutions of the problem with regard to the 'best before' date label could be provision of information to companies about the legal possibilities with regard to amendment of the 'best before' date, in order to encourage them to start placing new 'best before' dates on products if the original dates have passed, and the products still meet the conditions for sale (Waarts et al. 2011). Results of the survey in the UK showed that communication campaigns in the form of leaflet drop, promotional letters, face-to-face engagement were most effective in reducing overall residual waste and its food waste component (Wells et al. 2011).

4.1.2 Guidelines

Other informative tools are consumer guides and handbooks, describing such issues as avoidance of wasting food by shopping according to the daily needs, better planning and shopping patterns, as well as side effects of impulsive food shopping and consumption patterns (Gustavsson et al. 2011). Such types of information can be prepared and disseminated by stores in-store, through their websites and magazines (BIO Intelligence Service 2010), by public authorities, industry associations, and NGOs. Furthermore, such initiatives can include the provision of storage advice, information on labelling, distribution of tips for leftover cooking and 'packaging laboratory: keep it fresh' tests to identify what type of packaging can extend the life of specific fruit and vegetables (BIO Intelligence Service 2010).

4.1.3 Education

Education programmes on waste and date labels will help consumers to better understand the meaning of the terms and will increase their ability to judge the quality of produce and lead to reduction of the discard of food items (Kantor et al. 1997; Foresight 2010; Lin et al. 2013). In addition, education will improve consumers' understanding of the sociological dimensions of food consumption in different cultural, social and economic settings (Beddington et al. 2012), and support the change in the way they value food (Lin et al. 2013). Changes of social norms should lead to a point, where wasting food is considered unacceptable by a society (Foresight 2010). Moreover, education will help meal planners determine appropriate portion sizes, distinguish between spoiled and safe food, and better utilize leftovers (Kantor et al. 1997).

Training programmes include teaching of food waste prevention skills, workshops for consumers on waste-free cooking (BIO Intelligence Service 2010), usage of leftovers, as well as teaching staff about the impact of food waste and the methods of its avoidance during their work.

Guidance and training on the practical implementation of food donation are also needed because in some cases social institutions do not have experience in handling such types of food (BIO Intelligence Service et al. 2011).

In addition to the aforementioned factors that consumers should be aware of, it is important to educate them about the environmental impacts of food waste, and its contribution to carbon emissions. People also need to know about today's resource-constrained world (Lin et al. 2013).

The WRAP experts also see local infrastructure, such as separate food waste collections, as an additional way to affect consumer behaviour i.e. it makes them aware of the amount of food they throw away and change their purchasing, consumption and disposal habits (WRAP 2012).

4.2 Food Recovery and Redistribution

'Food recovery' and 'Food redistribution' could be considered as interchangeable terms. 'Food recovery' is a collection, or recovery, of wholesome food from farmers' fields, retail stores, or foodservice establishments for distribution to the poor and hungry (Kantor et al. 1997). Whereas, 'food redistribution' is voluntarily giving away food that otherwise would be lost or wasted to recipients e.g. charitable organisations, which then redistribute the food to those who need it (Lipinski et al. 2013). This approach applies at the production stage with unharvested crops, the manufacturing stage with overproduced products, and the distribution and retail stages with food left unsold at stores and markets (Lipinski et al. 2013). The following food types are considered for recovery for human consumption: (Kantor et al. 1997)

- Edible crops remaining in farmers' fields after harvest;
- Produce rejected because of market 'cosmetics' standards (e.g. blemishes, misshapen);
- Unsold fresh produce from wholesalers and farmers' markets;
- Surplus perishable food from food-service establishments e.g. restaurants, cafeterias, caterers, grocery stores;
- Packaged foods from grocery stores, including overstocked items, dented cans, and seasonal items

USDA, in its 'A Citizen's Guide to Food Recovery', distinguished four types of recovery activities: (Kantor et al. 1997)

- **field gleaning**—the collection of crops from farmers' fields that have already been mechanically harvested or on fields where it is not economically profitable to harvest;
- **perishable food rescue or salvage**—the collection of perishable produce from wholesale and retail sources such as supermarkets;
- **food rescue**—the collection of prepared foods from the foodservice industry, including restaurants, hotels, and caterers;
- **non-perishable food collection**—the collection of processed foods with longer shelf lives

There are different types of charitable organisations involved in such activities. One of them is food banks. The first banks were established in the late 1960s in the US. The movement has spread around the world under various different names, but following the same goal. Food banks collect food from a variety of sources, save it in a warehouse and distribute it to families and individuals in need through local social welfare organisations. The organisations served by food banks and food rescue programmes include community centres, shelters, soup kitchens, food pantries, childcare centres and senior programmes (Schneider 2008). Most of the European food banks are joined with the European Federation of Food Banks (FEBA), which was established in 1986 and supports the creation and development of food banks in European countries. Today, 253 European food banks are members of the organisation (FEBA 2013b). In 2010, according to FEBA, 240 food banks in 21 countries distributed 146 thousand tonnes of food products that were intended to be destroyed (FEBA 2013a). In the same year, according to the French Federation of Food Banks annual report, 98,780 tonnes of food were collected by only 79 food banks in France (Viel 2011). The French food banks cooperate with various actors of the food supply chain such as retail, food industry and agricultural cooperatives sectors as well as with consumers. In 2010, the retail sector 'donated' about 28 % of the total amount of food products received by the banks. It included edible products removed from the shelves before the deadline for consumption or released from supply platforms. The share of the total amount of food products received from the food industry sector and consumers constituted 21 and 14 % respectively (Viel 2011). In 2008, the Italian food redistribution programme 'Name 'amiat'' recovered over 81 tonnes of food, main types of which were bread and fruits (BIO Intelligence Service 2010).

In the UK, products which do not meet marketing standards for aesthetic reasons are redistributed to charitable organisations such as FareShare (BIO Intelligence Service et al. 2011). The organisation promotes the message 'No Good Food Should Be Wasted' and supports communities to relieve food poverty by providing surplus of food products from the food and drink industry to organisations working with disadvantaged people in the community (FareShare 2013).

However, food donation and following redistribution as a method of food waste reduction has a number of obstacles due to a number of transportation, legal, and economic factors (Lipinski et al. 2013). One of the main legal barriers is the issue of product liability. According to the law, prospective food donors are responsible for any possible 'health consequences of food' that recipients could suffer from. Therefore, one of the conditions for food donation is that potential food donors must be assured that personnel of food recovery programmes or in charitable organisations are trained in safe handling and storage of donated foods (Food Recovery Committee 2007).

From an economic perspective it would be inexpedient for a farmer to incur the labour, logistical, and transportation costs to donate food that was not harvested (Lipinski et al. 2013). Additional financial obstacles are costs of storing and packaging donated foods, securing labour whether paid or volunteer and their training in safe food handling and preparation methods (Kantor et al. 1997). However, despite these obstacles, experts believe that food redistribution is one of the efficient strategies to prevent food waste (BIO Intelligence Service et al. 2011).

4.3 Legislation—Governmental Interventions

Governments have a large number of tools to intervene in different areas of the food supply chain that will have a significant impact on its shaping. These include financial instruments, legal sanctions, regulation of claims, labels, guidance and etc. The application of one or more of these instruments influences on producers incomes, resource and environmental protection, consumer protection, wasteful practices of retailers, consumer preferences and purchasing patterns, the availability of foods and public awareness (Parfitt and Barthel 2011; Sonigo et al. 2012;

Institution of Mechanical Engineers 2013). Moreover, waste regulations play an important role in governing the options for owning food waste streams by assigning the highest possible extracting value to food waste. Successful examples of regulatory action include taxes on landfill, incentives for renewable energy, and standards for digestate (Ellen MacArthur Foundation 2013).

Today the European 'food waste relating legislation' is introduced by a variety of regulations and legislative documents some of which are perceived as obstacles; others as tools for food waste reduction. The list includes the following: (Waarts et al. 2011)

- European marketing standards
- Contamination in food
- Import control
- Phytosanitary policy
- Novel food
- Cooling and freezing meat
- Hygiene rules and product liability
- The provision of food information
- Norms and quotas in fisheries
- The use of animal by-products

Previously, the regulatory measures for food waste treatment were centred lower in the waste hierarchy (BIO Intelligence Service 2010). However, in more recent years the situation has changed and the EU has started giving more attention to the problem of food waste and initiated amendments of the aforementioned regulations and new initiatives. For instance, in 2012, the European Parliament called on the European Commission to encourage the discounted sale of food close to its expiry date (Ellen MacArthur Foundation 2013). Nevertheless, an analysis of the results showed that changes in legislation such as the phasing out of regulations or relaxed regulatory constraints on the aesthetic requirements for many fruits and vegetables (BIO Intelligence Service 2010) require time for business to take advantage of new opportunities (Waarts et al. 2011) and for entire sectors to adjust to new regulations (Redlingshöfer and Soyeux 2012). Therefore, in its guidelines for the development of food waste prevention programmes, the EU targets food manufacturing and processing, food retailing and distribution, food services and households sectors by stressing behavioural change and sectorial based approaches (Lin et al. 2013).

In general, most of the causes of food waste generation could be resolved through policies and regulations. For example, food date confusion can be reduced by implementation of policies or provision of guidance regarding dates appear on packages, including the information needed to understand these dates (Lipinski et al. 2013). An impact assessment analysis of possible policy options was conducted in the frame of 'Preparatory study on food waste across EU 27'(BIO Intelligence Service 2010). The results showed that the policy on the date labelling coherence, which involves a harmonisation of date labels on food products at the EU level via a requirement in the Food Information Regulation, and provides consistent messages to consumers on food safety, quality and optimum storage conditions, would lead up to 20 % of avoidable food waste reduction across Europe (BIO Intelligence Service 2010). During the study in the Netherlands, the chain actors pointed out that amendments in the Maximum Residue Levels (MRLs) level (i.e. higher level of MRLs) which actually reflects food safety for humans would lead to much less food being rejected or destroyed. However, it is important to emphasize that these amendments need not yield any risk in terms of food safety (Waarts et al. 2011).

The obstacle of product liability that prevents food donations could be overcome by 'Good Samaritan' laws. The law limits the liability of donors in case redistributed food unexpectedly turns out to be somehow harmful to the consumer. It assures food donors that they will not be penalised for redistributions made in good faith (Lipinski et al. 2013). In the United States the Bill Emerson Good Samaritan Act was passed in 1996. It protects food donors from civil and criminal liability if the product they redistributed in good faith to a charitable organization later causes harm to the needy recipient (Lipinski et al. 2013). The existence of such laws very effectively facilitates a redistribution process of food products to non-profit organizations and food rescue programmes (Food Recovery Committee 2007; Foresight 2010; Value Chain Management Centre 2012).

One of the latest legal steps taken by the European Parliament is the resolution on 'how to avoid food wastage: strategies for a more efficient food chain in the EU' (European Parliament Resolution (2011/2175 (INI)) 2012), where it urges the Council and the Commission to designate 2014 the European Year against Food Waste, as a key information and awareness-raising initiative for European citizens. The document also asks to take practical measures towards halving food waste by 2025, promote awareness-raising campaigns to inform public of the value of food and agricultural produce and reduce consumers' uncertainty regarding food edibility by clarifying the meaning of the date labels (European Parliament Resolution (2011/2175 (INI)) 2012).

In most of the cases a single policy does not have a significant impact on the food waste reduction process as whole, because it affects only one aspect e.g. social, economic or financial, therefore its successful implementation directly depends on other legislative changes.

4.4 Economic Incentives/Financial Instruments

The importance and predominance of a financial consideration during a decision making process, regardless of the extent and significance of a problem, make economic incentives one of the key measures in food waste reduction strategies. Financial instruments can be considered as levers that motivate actors of the food supply chain to change their behaviour with regard to an amount of food waste generated. However, the analysis of the current situation and experts' opinions talk about an insufficient number of this type of incentives. In general these instruments can be divided into two categories. The first category is negative financial stimuli which ‘enforce’ actors to bear additional costs; the second category is positive financial stimuli that offer additional economic benefits in the case of food waste reduction.

4.4.1 Negative

4.4.1.1 Internalisation of Costs

According to the effect of price-elasticity of demand, raising the cost of production and consumption of a resource will reduce its overall consumption (Maxwell et al. 2011) and will make to value it more. The same principle can be applied to food waste. Higher costs will lead to reduction of the amount disposed. This could be done through bonus-malus schemes (in the case of food waste the cost of disposal increase exponentially as the total quantity of food discarded) (Maxwell et al. 2011), or by internalising environmental impacts of waste (to attach a full account of the cost of food waste, e.g. include the carbon cost of waste) which will disclose full cost to consumers (Foresight 2010).

4.4.1.2 ‘Polluter Pays’ Principle

Another instrument is the ‘polluter pays’ principle, which would reduce food waste through enforcing those who are responsible for the waste (e.g. retailer/supplier relationships, where surplus is returned to the supplier) to pay its price (Foresight 2010).

4.4.1.3 Taxes and Charges

A further method is taxes and/or charges. It might include taxes and charges on products, on landfill and incineration, as well as differential charging for household waste (e.g. user fees, pay as you throw) (Arcadis et al. 2010). The instrument would effectively place responsibility and control throughout the supply chain (Foresight 2010), and making the charges as big as possible will lead to reduction of food waste (Waarts et al. 2011).

4.4.1.4 Rise of Food Prices

In one way or another the aforementioned instruments raise the food prices. This fact can be considered as a positive driver/stimulus for food waste reduction, because it has been argued that food prices are currently too low (Foresight 2011).

However, the issue of the food prices also touches upon social aspects and, first of all, might compromise the issue of food security for the poor (Foresight 2011). Therefore, all instruments that lead to an increase of the food prices for consumers must be considered very carefully and separately for each country. The measures aimed at an increase of the prices for other actors of the food supply chain must be implemented together with instruments that prevent the actors to pass these additional costs to a consumer.

4.4.2 *Positive*

Potential financial benefits of food waste reduction include a reduction of operating costs and increased revenue as a result of elimination of disposal costs, improved logistical and management practices. Additional revenue could come from improved ethical/social image of a company (Kantor et al. 1997; Value Chain Management Centre 2012). Another possible financial incentive for the food industry to minimise their food waste could be bigger profit margins in the sector, which increases options for dealing with the waste. In the United States the business environment is favourable towards food redistribution, because there are tax incentives to do so and together with legislation it also helps to improve corporate image (Hodges et al. 2011). Additional financial incentives for producers are subsidies, tax credits which will support innovations and R&D in the area, and funding of prevention programmes that focused on food waste reduction (Arcadis et al. 2010). In turn, consumers could be motivated to reduce their food waste by such economic argument as saving money by managing better food planning and purchasing (Foresight 2010). It could be spread with help of information and public awareness campaigns. Out-dated food products that are considered by most of the actors of the food supply chain as food waste still have an economic value/potential. It was proved by an innovative private-sector approach implemented in the UK. The main idea is to avoid food waste via resale (BIO Intelligence Service 2010). There are a number of online retailers such as ‘Approved Food and Drink Company’ and ‘Food Bargains’ which are specialized in selling approved out of date food through their website. ‘Approved Food and Drink Company’ sells dry food products that are near or past their ‘best before’ date at a discounted rate (BIO Intelligence Service 2010), whereas Food Bargains focuses on full date, clearance, short dated and out of date food and drink (Food Bargains 2010).

4.5 Forecasting and Correct Inventory Management/Planning

The issue of stock planning, to the same extent, is a problem for households and companies. The process requires consideration of many different aspects e.g.

projections of future demand of a product, seasonal fluctuations, current stock situation, in the case of households food whims of family members, special occasions and 'fullness' of a fridge.

In order to improve the accuracy of their forecasting demand techniques and thus to reduce food waste retailers could employ information technologies (Escaler and Teng 2011), evaluate marketing practices and communicate on how food should be stored/handled to staff and shoppers (Value Chain Management Centre 2012). It also requires participation and cooperation of all actors of the food supply chain. The results of the research in several EU countries showed that more exact and timely data about storage quantities from supermarkets leads to reduction of food waste (Ujhelyi 2013). However, the higher the accuracy level of demanded quantities is, the more extensive and detailed data regarding current stock, history sales, date labelling and etc. are required. Therefore, technical solutions have become an integral part of stock management practices. Technology is considered as an additional instrument to support and improve a variety of methods of food waste reduction. However, it is also important to note that technical solutions can be effective only when other elements of the food supply chain are effective (Lipinski et al. 2013).

Optimisation of the forecasting process can be implemented by using such technological innovations as

Automated replenishment systems help to reduce to a minimum the gap between predicted and actual sales (Weber et al. 2011). The system is based on sophisticated forecast in order to avoid overstocks, and it is crucial that a retailer has accurate inventory data. Buying and/or ordering decisions are made based on current stock levels, as well as daily selling patterns, product life-cycle, seasonality, projected waste, target service levels and inventory availability (Weber et al. 2011).

Real-time inventory management systems help to maintain accurate stock levels at all times. The system monitors the stock level during the course of the day (Weber et al. 2011).

'Smart shelves' indicate when stock levels are low and when products are nearing their 'sell-by' dates (Weber et al. 2011).

Radio Frequency Identification technology (RFID). Indicators on the packaging, based on this technology, measures storage conditions over time and reflect the actual shelf life, which helps in adjusting the 'best before' date to match the true shelf life of a product (Waarts et al. 2011).

Supermarket **loyalty cards** track product sales, by using information about shoppers and their purchases, and it helps to anticipate demand (Weber et al. 2011; BIO Intelligence Service et al. 2010). However, usually this type of programs asks customers to sign up and such lack of anonymity might not be welcomed by many customers.

Online shopping in advance allows stores to plan better, reduce overstocking and avoid compulsive purchasing. However, despite the fact that consumer research indicates that on-line shoppers are more able to plan their shopping and their meals and are less prone to being attracted to retail promotions, there are sceptics that think that such behaviour might still lead to waste. Internet offers the possibility of cutting out the middle player—the retailer. For instance, 30 years ago retailers had much less power in the UK than they do today (Foresight 2010). Food service providers might offer an option to their customers to make their meal reservations/ orders through an online system, it would help companies to get more precise data about future demand and better plan their menu.

Technology available in homes that will help households to reduce amount of food waste over next decades will be **smart fridges** that 'tell' consumers what they (fridges) contain and automatically reorder key food items (Foresight 2010). Moreover, advanced devices are able to monitor food purchase and waste, and provide metrics and indicators of the efficiency of food use at the level of an individual household (Foresight 2011).

4.6 Packaging

Packaging could also be considered as one of the tools for food waste prevention. It extends shelf and storage life of a product (e.g. lightweight packaging, 'active and intelligent packaging') (UK Government 2010; Foresight 2011; Ellen MacArthur Foundation 2013). Moreover, packaging offers protection from spoiling, increases transportability of food products and provides information to a consumer (Sonigo et al. 2012). There are initiatives promoting an addition of new functionalities to sealing and reclosure systems to prevent products drying out, hardening or spilling (Sonigo et al. 2012; Redlingshöfer and Soyeux 2012).

4.7 Labelling

Changes and/or improvements of the current food labelling system would also have a significant impact on the reduction of food waste. By providing different types of information, labels influence the attitude of consumers and purchasers (Arcadis et al. 2010). All labels need to be smart and simple (Foresight 2010). One of the possibilities to avoid confusion between information relevant for retailers and manufacturers, or only for consumers is to make this part of the label content hidden i.e. it could be scanned or read only by a manufacturer and retailer. This would prevent consumers from misinterpreting a date on a package and throwing the item away prematurely. This approach could be applied to non-perishable foods and will require a small shift in packaging manufacturing processes (Lipinski et al. 2013).

In order to increase consumers understanding of the environmental costs associated with each food product, experts recommend to add to the label content environmental information regarding carbon values and wastage rates (Foresight 2010).

Starting from the year 2014, the deployment of the new barcode system **GS1 Databar** will enable manufacturers to store and retailers to get information about serial and lot numbers, and expiry dates. The system will become an alternative to the Global Trade Item Number (GTIN) used today (Weber et al. 2011) and help to improve stock management practices.

4.8 Companies Initiatives

There are a number of additional measures for food waste reduction which can be implemented by companies on a voluntary basis at different stages of the food supply chain. Companies can reduce food waste by imposing less stringent requirements (private marketing standards) of the shape of fruit and vegetables and by adjusting their logistical processes accordingly (Waarts et al. 2011). WRAP recommends retailers to provide freezing tips, maximising shelf-life, at the shelf edge alongside the promotional offer, flexible meal plans and encourage batch cooking/freezing (e.g. cook once eat twice) (WRAP 2011). Some retailers mark down the price of goods that are about to reach their 'use by' date at the end of each day (Foresight 2010). Others, like 'Sainsbury's' and 'Morrison's', the third- and fourth-largest grocery retailers in the UK respectively have organised waste reduction campaigns. These campaigns highlight the issue of food waste for consumers and provide them with tips for reducing waste. Customers are reached through in-store displays, pamphlets, and websites that contain recipes, storage tips, and information on freshness and shelf lives of food products (Lipinski et al. 2013).

The 'Co-operative Group' retailer has begun printing tips for improving food storage and lengthening shelf-life for fruits and vegetables directly onto the plastic produce bags in which customers place their purchases (Lipinski et al. 2013).

The retail chain 'Albert Heijn' has developed food waste action program regarding both retailer and consumer sides. The following activities are implemented on the retailers' side: (Arcadis et al. 2010)

- **monitoring**: a team of waste specialists monitors daily the sales of (most fresh) products that helps to reduce the waste amounts of low-performing products;
- **logistics**: smart logistic chain guaranteeing that products that decrease in sales are supplied in smaller amounts;
- **marking** down products which are close to being wasted.

With regard to consumers the company is planning to

- improve the clarity and consistency of date labelling and storage guidance;
- help consumers to know what they need to buy, and how much;
- let consumers take full advantage of special offers by knowing how to manage the extra food offered through these promotions (e.g. recipes);
- optimise packaging.

In order to reduce amounts of unplanned food bought only because of promotion discounts, the grocery retailer 'Tesco' launched a 'buy One Get One Free Later' initiative to allow customers buying perishable goods to collect their free item later. The programme works through a voucher system. Products included in the initiative are those which are considered 'short-code life perishable products' with short sell dates such as yoghurts, salads, vegetables and cheese (BIO Intelligence Service 2010).

Changes in an operation of the catering system of the Hvidovre Hospital, in Denmark, such as offering anytime 'a la carte' order options to patients have helped the hospital avoid 40 tonnes of food waste per year, and encourage portion management (BIO Intelligence Service 2010). The issue of portion size is also crucial for food service providers e.g. restaurants, caterers. One approach is to offer various portion sizes at different prices. It would allow customers with smaller appetites to order a smaller meal and presumably leave less of it behind, while also lowering preparation costs for a restaurant (Lipinski et al. 2013). In a buffet or cafeteria-style food service environment, companies could post informational signs reminding customers to take only as much food for which they have the appetite. Another possibility is not to offer cafeteria-style trays, customers will carry the food they purchase on plates, it also prevents 'hoarding' behaviour (Lipinski et al. 2013). The next possible approach is to replace 'all-you-can-eat' options from buffets with 'pay-by-weight' systems in which the weight of the plate of food determines the cost of the meal (Lipinski et al. 2013). Results of an experiment with a 'trayless cafeteria' at Grand Valley State University (GVSU) in the United States in 2007, showed that an amount of food that has been discarded reduced by almost 13 metric tonnes—about 25 kg per person annually (Lipinski et al. 2013). The problem of portion or to be more precisely of the single-packaging size of a product in supermarkets can be solved through self-dispensing systems. Goods are taken to stores in bulk packaging and sold without primary consumer packaging. It allows consumers to buy as much as they actually need.

The impact of single company initiatives is not as significant as results of collaboration of whole sector or/and between sectors. Such cooperation can be achieved through voluntary agreements, which also can be considered as an alternative to regulations (Foresight 2010) (Parfitt and Barthel 2011). One example of such agreements is the Courtauld Commitment in the UK. In 2005 over 40 major retailers, brand owners, manufacturers and suppliers have signed the document and committed to reduce both post-consumer packaging and post-consumer food waste through innovative packaging and optimal choice of volume of the product, in-store guidance and consumer campaigns (e.g. Love Food Hate Waste) (Arcadis et al. 2010).

Another example is the Joint Food Wastage Declaration 'Every Crumb Counts' (The Joint Food Wastage Declaration "Every Crumb Counts" 2013). The co-signers commit to reduce food waste throughout the food chain and contribute to halving EU edible food waste by 2020. They also call on all stakeholders involved in the food chain from farm to fork and beyond to take further action to prevent and reduce edible food waste on a European and global scale (The Joint Food Wastage Declaration "Every Crumb Counts" 2013). Among co-signers are: (The Joint Food Wastage Declaration "Every Crumb Counts" 2013)

- European Organisation for Packaging and the Environment
- European Potato Trade Association
- European Federation of Food Banks
- European Food and Drink Industry
- Sustainable Restaurants Association
- European Fresh Produce Association
- European Contract Catering Industry, etc.

There are different types of voluntary agreements that can have impact on most stages of the product life cycle, depending on the parties that are involved. These could be agreements between government and the industry, the industry and consumer organisations, government and the distribution sector, governments and notified bodies in the frame of extended producer responsibility schemes, consumer agreements and etc. (Arcadis et al. 2010).

Methods of food waste reduction discussed below belong to the lower part of the food waste recovery hierarchy, in other words, from this point and on food is treated as waste. The main goal of these activities is to divert food waste from landfills. One of the requirements of successful realization of these methods is separate collection of the food waste stream.

4.9 Separate Collection of Food Waste

Keeping the food waste stream separate from other types of bio-waste e.g. garden waste would have a significant impact on the quantity and quality of material remaining in the refuse stream and costs of its disposal (Lamb and Fountain 2010). Furthermore, it will allow optimisation of the processing routes for each bio-waste element (Arcadis 2010) and help to avoid toxic substances in food waste composition (Ellen MacArthur Foundation 2013). The final product with a higher quality can draw higher value and be utilised in a greater variety of end markets (Lamb and Fountain 2010). Moreover, separate collection contributes towards improving recycling rates, reducing greenhouse gas emissions and the environmental impacts associated with landfill (toxicity in leachate, landfill gas emissions etc.) (WRAP 2009). Separation and measurement of food waste will also make staff (and sometimes customers) aware more of the quantity of food waste they are generating, provoking efforts to reduce this (BIO Intelligence Service et al. 2010). Types of a food waste collection system define methods of its treatment. For instance, composting facilities that accept food waste combined with garden waste may find it harder to control ratios of nutrients in the feedstock, which has implications for end product quality (WRAP 2009). It is also important to note that successful implementation of the separate food waste collection system directly depends on positive attitudes and high participations rate by the public towards the activity (Karim Ghani et al. 2013), commitment to recycling, cultural influences on cooking habits, home composting rates, amount of food left in packaging and etc. (WRAP 2009).

4.10 Alternative Use

The stage when food waste cannot be redirected for human consumption, does not have to be seen as a last one before food waste disposal. It still can be considered as a valuable resource because of its huge potential as a source for energy, nutrients, or carbon (Ellen MacArthur Foundation 2013). This is the second best option of using food (Schneider 2008). The results of the study conducted by the Ellen MacArthur Foundation showed that an application of the circular economy, which undermines assignment of a value to waste by using it for other purposes e.g. biogas generation, animal feed and return of nutrients to agricultural soils, would globally divert up to 272 million tonnes of food waste from landfill annually (Ellen MacArthur Foundation 2013). Only in the UK a possible income, which also includes revenue from feed-in-tariff and avoided landfill fees, would amount to USD 1.5 billion (Ellen MacArthur Foundation 2013). In addition, such alternative use of food waste would make a positive contribution to the global food and energy balance (Foresight 2011).

4.10.1 Energy Recovery

There are a number of technologies on the market for biological treatment of bio-waste that could be used for energy recovery. The most environmentally preferable is anaerobic digestion (UK Department of Energy and Climate Change & Defra 2011).

4.10.1.1 Anaerobic Digestion (AD)

Anaerobic digestion (AD) is a process in which microbes digest organic material in the absence of oxygen. The process creates two distinct products: biogas and a liquid or solid residue, the digestate (Ellen MacArthur Foundation 2013). A gas can be burnt to generate electricity and heat, further processed into biofuel, or injected into the gas grid (UK Government 2010; UK Department of Energy and Climate Change & Defra 2011). In Europe the cumulative capacity of more than 200 anaerobic digestion plants amounts to 7,750 thousand tonnes per year (Baere and Mattheeuws 2012). The current leaders among the European countries using AD plants are Germany, Spain and France (Fig. 4.1).

In addition to the reduced amount of waste sent to landfill, anaerobic digestion has a number of economic and environmental advantages. The construction of AD facilities can be comparatively fast and relatively inexpensive in comparison with some other waste management technologies (UK Department of Energy and Climate Change & Defra 2011). Estimations made by the experts showed that processing food waste with anaerobic digestion could create additional profit, which

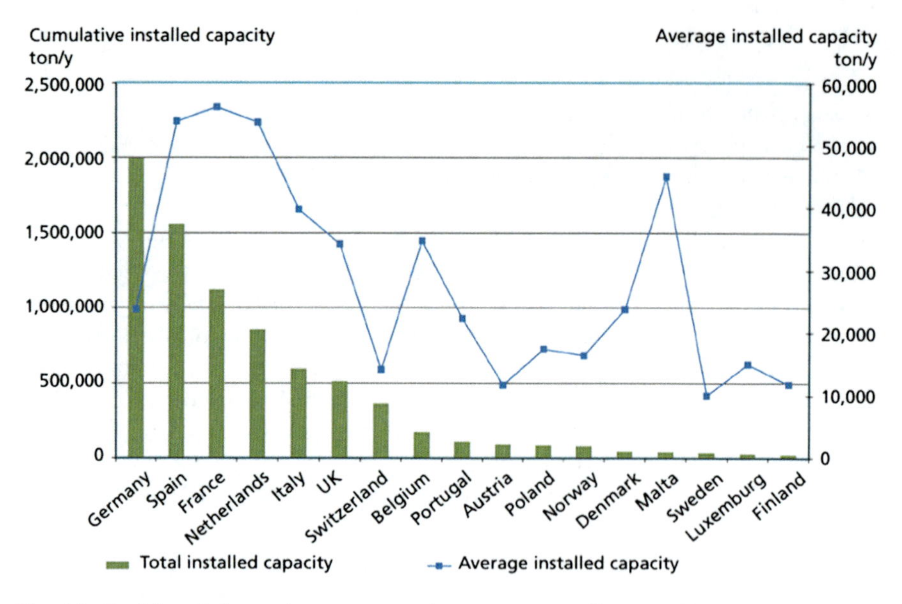

Fig. 4.1 Total installed capacity per country (tonnes per year) (Baere and Mattheeuws 2012)

includes avoiding the cost of landfill, sales of electricity, heat and fertilizers. The study in the UK showed that an operating profit would amount up to USD 172 per tonne of food waste (Fig. 4.2). However, the calculations were made with the following assumptions: (Ellen MacArthur Foundation 2013)

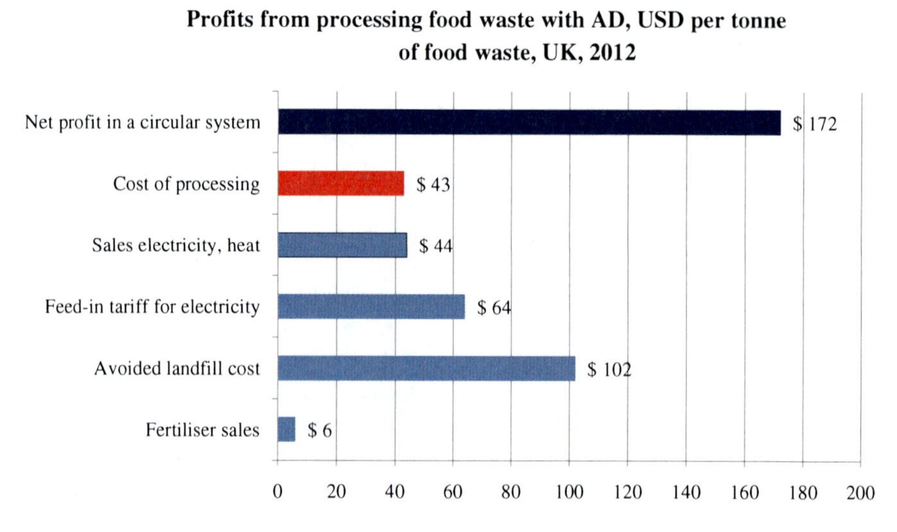

Fig. 4.2 Processing food waste with anaerobic digestion instead of landfilling (Ellen MacArthur Foundation 2013)

- Large-scale AD rollout will reap benefits from bulk orders and operational efficiency gains;
- New bacteria strains will increase throughout;
- Digestate will be sold as a full equivalent to mineral fertiliser (vs. none today) because of increased uptake and development of new dewatering technologies

Environmental advantages of AD, amongst others, are avoiding landfill emissions, returning organic material and fertilising nutrients to the soil (Ellcn MacArthur Foundation 2013).

4.10.1.2 Incineration

Despite the fact that incineration is more expensive process than anaerobic digestion (Ellen MacArthur Foundation 2013), it is still seen as an alternative method of energy recovery from food waste. It is recommended, especially, when food waste contains animal by-products in order to avoid potential risk to animals, the public and the environment (Defra 2013). Waste to energy plants could be used for producing electricity, steam and heating (Bogner et al. 2007; European Commission 2010). In 2011, the USA combusted about 12 % of their MSW for energy recovery (U.S. EPA 2013b). In Europe thermal treatment of waste is covered by the EU Waste Incineration Directive (2000/76/EC) (Directive 2000/76/EC 2000). Today, the Waste-to-Energy sector contributes to 23 % of MSW management in Europe (CEWEP 2013). According to the Eurostat data, the percentage of waste treated by the incineration with energy recovery method in some European countries is above 20 % (Fig. 4.3), out of which, the percentage of animal, mixed food waste and vegetal wastes reaches 12 % (Fig. 4.4).

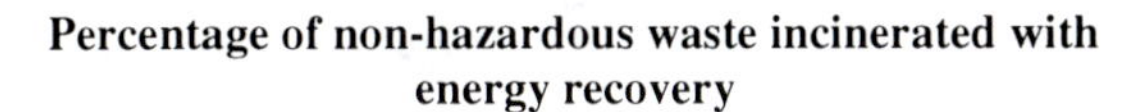

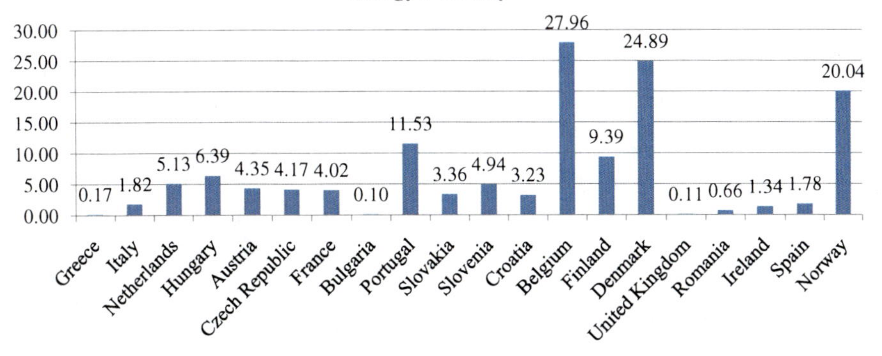

Fig. 4.3 Percentage of non-hazardous waste incinerated with energy recovery in the European countries in 2010 (Eurostat 2013a, b)

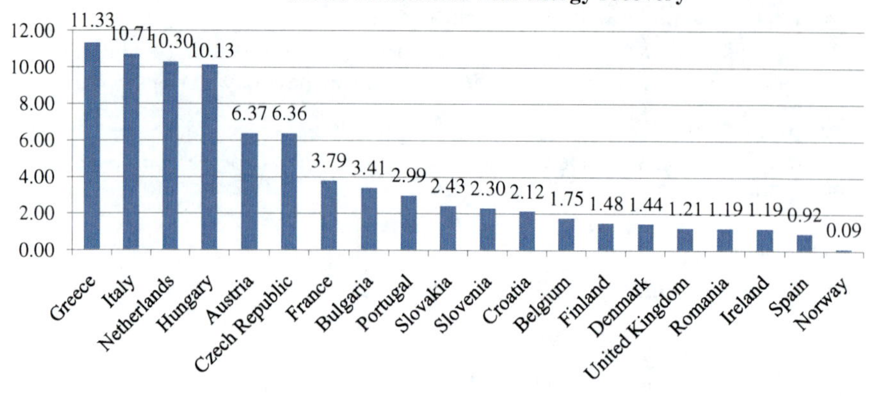

Fig. 4.4 Percentage of animal, mixed food waste and vegetal wastes incinerated with energy recovery in the European countries in 2010 (Eurostat 2013a, b)

4.10.2 Novel Added-Value Materials/Products

4.10.2.1 Chemicals and Fertilizers

Today, in a world of limited resources and the increasing demand for chemicals and fertilisers, food waste could be considered as an additional source for their production (Lin et al. 2013). Food waste is rich with valuable soil nutrients nitrogen (N), phosphorus (P), and potassium (K). These nutrients could be used as a part replacement for mineral fertiliser, soil improvement for farms, which is preferable from a nutrient and soil structure perspective, or as a source of chemicals (Sonigo et al. 2012; Ellen MacArthur Foundation 2013; Pfaltzgra et al. 2013). Another replacement for mineral fertilisers is the digestate from the AD process, which is a nitrogen-rich and also could be used for agriculture as a renewable fertiliser or soil conditioner (UK Government 2010; UK Department of Energy and Climate Change & Defra 2011; Ellen MacArthur Foundation 2013).

Moreover, food waste could be a renewable feedstock for the chemical industry. Table 4.1 includes the list of products and materials that could be produced by different branches of the chemical industry from food waste.

4.10.2.2 Compost

Another end-product from food waste that could be used as an organic fertiliser is compost. **Composting** is a biological process in the presence of oxygen during which microorganisms (e.g. bacteria and fungi), insects, snails, and earthworms break down organic materials into a soil-like material called compost (Ellen

Table 4.1 Sectors of the chemical industry that could benefit from the use of food supply

Sectors of the chemical industry	Products/materials
Speciality chemicals	*Methods of food waste reduction* Flavour and fragrances Antioxidants Adhesives Pharmaceuticals
Consumer chemicals	Home and personal care Coatings Food additives
Commodity chemicals	Chemical building blocks Synthetic fibers Fuel Plastics and rubbers

MacArthur Foundation 2013). It reduces the volume of water and kills pathogens while preserving more of the nutrients for use by crops (U.S. EPA 2012). Compost quality defines an area of its application e.g. agriculture, horticulture, soil stabilization and soil improvement (increased organic matter, higher water-holding capacity) (Bogner et al. 2007).

There are several types of composting:

In-vessel composting (IVC) is an industrial form of rapid composting under controlled conditions (Ellen MacArthur Foundation 2013). Waste is processed through large-scale enclosed composting plants, in the case of food waste such facilities must also be compliant with Animal By-Product Regulations (ABPR) (UK Department of Energy and Climate Change & Defra 2011).

Windrow composting usually relies on natural processes for air supply to the waste, although it may be artificially aerated. Windrows are turned to increase the porosity of the pile, and increase the homogeneity of the waste (European Commission Directorate-General for the Environment 2000). The process is suitable for fruit and vegetable waste, and all catering waste from households (Arcadis 2010). This method can accommodate large volumes of diverse wastes, including, grease, liquids, and animal by-products (such as fish and poultry wastes) (U.S. EPA 2013c).

Home composting scheme: biodegradable waste generated by householders is used to produce compost for use by an individual (European Commission Directorate-General for the Environment 2000). However, this method is not suitable for animal products or large quantities of food scraps (U.S. EPA 2013c).

The benefits of compost depend on a number of factors such as climatic conditions, soil quality, competition from other fertilisers (such as manure) and limitations on the amounts of nitrogen and phosphorus that can be applied to the land, which are varied across Member States (Arcadis 2010).

In 1995, more than 13 million tonnes of municipal waste were composted by the European countries. In 2008, this had accounted for 17 % of municipal waste

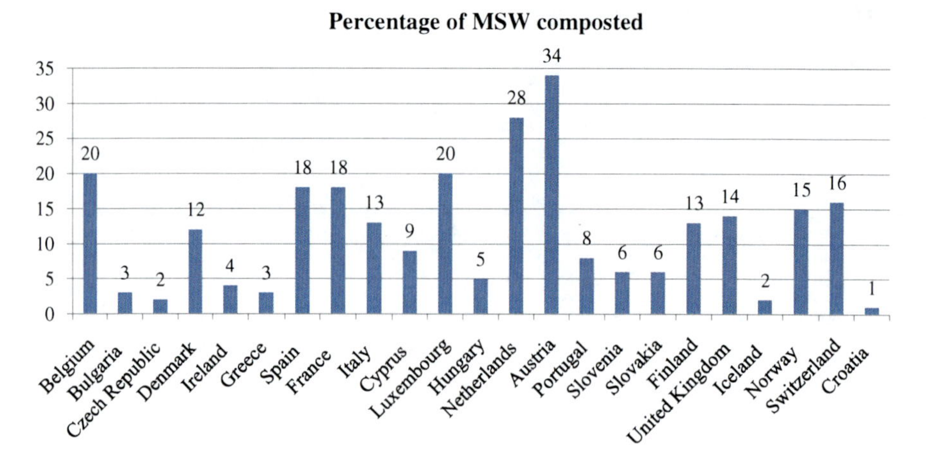

Fig. 4.5 Percentage of MSW composted per person in the European countries in 2011 (Eurostat 2013a, b)

(European Commission 2010). In 2011, 15 % of MSW was composted per person in EU-27 (Eurostat 2013b). In some countries this value reaches more than 30 % (Fig. 4.5).

4.10.2.3 Animal Feed

Another way to use food waste is for animal feed purposes, however, with a number of regulatory limitations (Lin et al. 2013). With new technology food waste could be used as a substitute for cereal in animal feed. According to experts, the 'released' cereal could feed 3 billion people and greatly reduce pressures on biodiversity and water resources (Nellemann et al. 2009). The main difficulties regarding use of food waste for animal feed purposes lie in its composition. Meat or animal materials might lead to the spread of disease and high concentration of some components can be harmful to livestock. Therefore, there are different examples of using and non-using food waste for animal feed around the world.

In order to decrease dependence on imported food, increase overall self-sufficiency in animal feed and avoid landfilling, processing food waste into animal feed is highly promoted in Asia (Ellen MacArthur Foundation 2013).

At the same time the UK prevents the use of food waste for animal feed from households, supermarkets, and the hospitality industry because it may contain meat (Ellen MacArthur Foundation 2013).

In the United States regulations vary from state to state. Some states ban food donation for animal feed while others regulate what food can be donated. The regulations may also require specific handling processes in order to donate to animal feedstock (U.S. EPA 2013a). One of the successful business examples in the US is 'Barthold Recycling and Roll-Off Services' company, which starting from

Table 4.2 Methods of food waste reduction

Methods	Implications
Public awareness raising/ education	• Informative tools have an impact on every stage of the product life cycle • Experts agree upon the necessity, importance and effectiveness of direct communication and awareness-raising among consumers • The main goal is to motivate or 'enforce' actors to change their behaviour regarding food • Inclusion of the food waste issue into national plans
Awareness campaigns and informativeness	• Address causes of household food waste: lack of awareness and knowledge on methods for avoiding food waste, date label confusion and inappropriate storage • Effective communication to households on the benefits of food waste recovery to maximise their participation in the process • Government campaigns to inform consumers and make them more aware of their wealth and the value of food • Provision of information to companies about the legal possibilities with regard to amendment of the 'best before' date, to encourage them to place new 'best before' dates on products if the original dates have passed, and the products still meet the conditions for sale • Communication campaigns in the form of leaflet drop, promotional letters, and face-to-face engagement are most effective in reducing food waste
Guidelines	• Consumer guides and handbooks, describing ways to avoid food waste: shopping according to the daily needs, better planning and shopping patterns, side effects of impulsive food shopping and consumption patterns • Include storage advices, information on labelling, tips for leftover cooking and 'packaging laboratory: keep it fresh' tests to identify what type of packaging can extend the life of specific fruit and vegetables • Are disseminated by stores, in-store, through their websites and magazines, by public authorities, industry associations, and NGOs
Education	• Education programs on waste and date labels to help better understand meaning of the terms • Education on the environmental impacts of food waste, its contribution to carbon emissions and today's resource-constrained world • Increase of consumers' ability to judge the quality of produce and lead to reduction of the discard of food items • Improvement of consumers' understanding of the sociological dimensions of food consumption in different cultural, social and economic programs: teaching of food waste prevention skills, workshops for consumers on waste-free cooking, usage of leftovers, teaching of staff about an impact of food waste and methods of its avoidance during their work

(continued)

Table 4.2 (continued)

Methods	Implications
	• Guidance and training on the practical implementation of food donation
Food recovery and redistribution	• The approach applies at the production, manufacturing, distribution and retail stages • Food types considered for recovery for human consumption: – Edible crops remaining in farmers' fields after harvest – Produce rejected because of market 'cosmetics' standards (e.g. blemishes, misshapen) – Unsold fresh produce from wholesalers and farmers' markets – Surplus of perishable food from restaurants, cafeterias, caterers, grocery stores and etc. – Packaged foods from grocery stores, including overstocked items, dented cans, and seasonal items • Types of recovery activities: – field gleaning – perishable food rescue or salvage – food rescue – non-perishable food collection • Charitable organisations involved: Food Banks, FEBA, FareShare—goal: to transfer surplus of food products from the food and drink industry to people in need
Legislation—governmental interventions	• Governments tools to intervene: financial instruments, legal sanctions, regulation of claims, labels, guidance and etc. • Waste regulations assign to food waste the highest possible extracting value • Successful regulatory actions: taxes on landfill, incentives for renewable energy, standards for digestate • European 'food waste relating legislation': – European marketing standards – Contamination in food – Import control – Phytosanitary policy – Novel food – Cooling and freezing meat – Hygiene rules and product liability – Provision of food information – Norms and quotas in fisheries – The use of animal by-products • The European Parliament called on the European Commission to encourage the discounted sale of food close to its expiry date • The policy on the date labelling coherence: a harmonisation of date labels on food products, provision of consistent messages to consumers on food safety, quality and optimum storage conditions

(continued)

Table 4.2 (continued)

Methods	Implications
	• Amendments in the MRLs (i.e. higher level of MRLs) lead to less food being rejected or destroyed • 'Good Samaritan' laws: limits the liability of donors in case if redistributed food unexpectedly turns out to be somehow harmful to a consumer. It assures food donors that they will not be penalized for redistributions made in good faith • The European Parliament: the resolution on 'how to avoid food wastage: strategies for a more efficient food chain in the EU' • The document asks to take practical measures towards halving food waste by 2025, promote awareness-raising campaigns to inform public of the value of food and agricultural produce and reduce consumers' uncertainty regarding food edibility by clarifying the meaning of the date labels
Economic incentives/financial Instruments	Levers to motivate actors of the food supply chain to change their behaviour with regard to an amount of food waste generated
Negative	
Internalisation of costs	• Higher costs lead to reduction of the amount of waste disposed • Bonus-malus schemes: the cost of disposal increase exponentially as the total quantity of food discarded • Internalisation of environmental impacts of waste: to attach a full account of the cost of food waste
'Polluter pays' principle	Enforcement of those who are responsible for the waste (e.g. retailer/supplier relationships where surplus is returned to the supplier) to pay its price
Taxes and charges	• Taxes and charges on products, landfill and incineration • Differential charging for household waste (e.g. user fees, pay as you throw) • Instruments effectively places responsibility and control throughout the supply chain
Rise of food prices	• A positive driver for food waste reduction, because food prices are currently too low • Might compromise the issue of food security for the poor • Application must be considered very carefully and separately for each country • The measures aimed at an increase of the prices must be implemented together with instruments that prevent to pass these additional costs to consumers
Positive	• Reduction of operating costs and increased revenue as a result of elimination of disposal costs, improved logistical and management practices • Additional revenue from improved ethical/social image of a company

(continued)

Table 4.2 (continued)

Methods	Implications
	• Bigger profit margins in the sector, increases options for dealing with the waste • Financial incentives for producers: subsidies, tax credits to support innovations and R&D in the area, funding of food waste prevention/reduction programs • Consumers: saving money by managing better food planning and purchasing • Innovative private-sector approach: resale of approved out of date food
Forecasting and correct inventory management/planning	• Requires consideration of: projections of future demand of a product, seasonal fluctuations, current stock situation, in the case of households: food whims of family members, special occasions and 'fullness' of a fridge • Employment of information technologies to improve an accuracy of forecasting demand techniques • Evaluation of marketing practices • Communication on how food should be stored/handled to staff and shoppers • Requires participation and cooperation of all actors of the food supply chain • Higher accuracy level of demanded quantities—more extensive and detailed data on current stock, history sales, date labelling and etc. • Technical solutions to optimise the forecasting process: – Automated replenishment systems – Real-time inventory management systems – 'Smart shelves' – Radio Frequency Identification technology – Supermarket loyalty cards – Online shopping – Meal reservations/orders through an online system – For households: smart fridges
Packaging	• Extends shelf and storage life of a product (e.g. lightweight packaging, 'active and intelligent packaging') • Protects from spoiling • Increases transportability of food products • Provides information to a consumer
Labelling	• Needs to be smart and simple • Different parts of a label available only for correspondent actors • Inclusion of environmental information regarding carbon values and wastage rates to the label content • New barcode system GS1 Databar: enables manufacturers to store and retailers to get information about serial and lot numbers, and expiry dates

(continued)

Table 4.2 (continued)

Methods	Implications
Companies initiatives	• Impose less stringent requirements (private marketing standards) on the shape of fruit and vegetables • Adjust logistical processes • Provide freezing tips, maximising shelf-life, at the shelf edge alongside the promotional offer, flexible meal plans and encourage batch cooking/freezing • Mark down the price of goods that are about to reach their 'use by' date at the end of each day • Organise waste reduction campaigns • Print tips for improving food storage and lengthening shelf-life for fruits and vegetables directly onto the plastic produce bags, in which customers place their purchases • Launch of a 'buy One Get One Free Later' initiative for 'short-code life perishable products': such as yoghurts, salads, vegetables and cheese • Offer various portion sizes at different prices, to encourage portion management • Replace 'all-you-can-eat' options from buffets with 'pay-by-weight' systems, in which the weight of the plate of food determines the cost of the meal • Voluntary agreements in the sector: – Courtauld Commitment in the UK – Joint Food Wastage Declaration 'Every Crumb Counts'
Separate collection of food waste	• Has a significant impact on the quantity and quality of material remaining in the refuse stream and costs of its disposal • Allows optimisation of the processing routes for each bio-waste element • Helps to avoid toxic substances in food waste composition • Contributes towards improving recycling rates, reducing greenhouse gas emissions and the environmental impacts associated with landfill • Makes staff and customers aware more of the quantity of food waste they are generating, provokes efforts to reduce it • Successful implementation of the separate food waste collection system directly depends on positive attitudes and high participations rate by the public towards the activity, commitment to recycling, cultural influences on cooking habits, home composting rates, amount of food left in packaging and etc.
Alternative use	• The second best option to using food waste: a source for energy, nutrients, or carbon • The main goal—to divert food waste from landfills

(continued)

Table 4.2 (continued)

Methods	Implications
Energy recovery	
Anaerobic digestion (AD)	• Creates products: biogas and a liquid or solid residue, the digestate • Gas: burnt to generate electricity and heat, further processed into biofuel, or injected into the gas grid • The current leaders among European countries using AD plants are Germany, Spain and France • The construction of AD facilities comparatively fast and relatively inexpensive • Processing food waste with AD creates additional profit: avoiding the cost of landfill, sales of electricity, heat and fertilizers • Environmental advantages: avoiding landfill emissions, returning organic material and fertilising nutrients to the soil
Incineration	• An alternative method of energy recovery from food waste • Recommended, when food waste contains animal by-products • Waste to energy plants used for producing electricity, steam and heating • In Europe thermal treatment of waste is covered by the EU Waste Incineration Directive 2000/76/EC
Novel added-value materials/products	
Chemicals and Fertilizers	• Additional source for production of chemicals and fertilisers • Food waste is rich with the valuable soil nutrients nitrogen (N), phosphorus (P), and potassium (K) • These nutrients are partial replacement for mineral fertiliser, soil improvement for farms • Nitrogen-rich digestate from the AD process used for agriculture as a renewable fertiliser or soil conditioner • Products and materials produced by the chemical industry from food waste: – Flavour and fragrances – Pharmaceuticals – Synthetic fibers – Plastics and rubbers
Compost	• Used as an organic fertiliser • Compost quality defines an area of its application e.g. agriculture, horticulture, soil stabilization and soil improvement • Types of composting: – In-vessel composting – Windrow composting

(continued)

Table 4.2 (continued)

Methods	Implications
	– Home composting scheme • The benefits of compost depend on: climatic conditions, soil quality, competition from other fertilisers, limitations on the amounts of nitrogen and phosphorus that can be applied to the land
Animal feed	• Used as a substitute for cereal in animal feed • The 'released' cereal could feed 3 billion people and greatly reduce pressures on biodiversity and water resources • Asia promotes the use of food waste to decrease dependence on imported food, increase overall self-sufficiency in animal feed and avoid landfilling • The UK prevents the use of food waste for animal feed from households, supermarkets, and the hospitality industry because it may contain meat • In the USA regulations vary from state to state

1988 has collected food from restaurants, hotels, schools, nursing homes, grocery stores and even large food processors to feed 3,800 pigs and 250 head of cattle on its 290-acre facility. Today, the company collects food scraps from about 400 commercial customers. In order to comply with the requirement of the Federal government and the State of Minnesota Department of Animal Health to process or cook food before feeding it to animals to kill harmful bacteria Barthold pioneered a method of cooking the food discards in the trucks after collection. Steam pipes are hooked to the truck for 20 min, increasing the temperature enough to kill potentially harmful bacteria (U.S. EPA 2006).

Table 4.2 summarises the main methods of food waste reduction, currently applied worldwide. Development and application of any of these methods require participation and cooperation of many of actors, only a small number of which are not responsible for food waste generation. The processes are very expensive and time consuming. The effectiveness and efficiency of outcomes of selected measure/s depends on numerous factors such as legislation, financial resources, available technologies as well as climatic conditions and prevailed economic sectors in a given country.

References

Arcadis et al. (2010). *Analysis of the evolution of waste reduction and the scope of waste prevention. Final Report.* Retrieved from http://ec.europa.eu/environment/waste/prevention/pdf/report_waste.pdf.

Beddington, J., et al. (2012). *Achieving food security in the face of climate change: Final report from the commission on sustainable agriculture and climate change. CGIAR Research*

Program on Climate Change, Agriculture and Food Security (CCAFS), Copenhagen. Retrieved from http://ccafs.cgiar.org/sites/default/files/assets/docs/climate_food_commission-final-mar2012.pdf.

BIO Intelligence Service. (2010). *Preparatory study on food waste across EU 27*. Retrieved from http://ec.europa.eu/environment/eussd/pdf/bio_foodwaste_report.pdf.

BIO Intelligence Service, Umweltbundesamt & Arcadis. (2011). *Guidelines on the preparation of food waste prevention programmes*. Retrieved from http://ec.europa.eu/environment/waste/prevention/pdf/prevention_guidelines.pdf.

Bogner, J., et al. (2007). *Waste management, in climate change 2007: Mitigation. Contribution of working group III to the fourth assessment report of the intergovernmental panel on climate change*, Cambridge, United Kingdom; New York, USA. Retrieved from http://www.ipcc.ch/pdf/assessment-report/ar4/wg3/ar4-wg3-chapter10.pdf.

Bridgwater, E., & Parfitt, J. (2009). *Evaluation of the WRAP separate food waste collection trials*. Retrieved from http://www.wrap.org.uk/sites/files/wrap/Evaluation_of_the_WRAP_FW_Collection_Trials_Update_June_2009.pdf.

CEWEP. (2013). *A decade of waste-to-energy in Europe*, Brussels. Retrieved from http://www.cewep.eu/m_1174.

De Baere, L., & Mattheeuws, B. (2012). Anaerobic digestion of the organic fraction of municipal solid waste in Europe. Status, experience and prospects. *Waste Management*, *3*, 517–526. Retrieved from http://www.ows.be/wp-content/uploads/2013/02/Anaerobic-digestion-of-the-organic-fraction-of-MSW-in-Europe.pdf.

Defra. (2013). Animal by-products: collection, storage and disposal. *Farming and the environment*. Retrieved Nov 5, 2013, from https://www.gov.uk/dealing-with-animal-by-products.

Directive 2000/76/EC. (2000). Directive 2000/76/EC of the European Parliament and of the council of 4 December 2000 on the incineration of waste. *Official Journal L, 332*, 91–111. Retrieved Dec 28, 2000 from http://eur-lex.europa.eu/LexUriServ/LexUriServ.do?uri=OJ:L:2000:332:0091:0111:EN:PDF.

Ellen MacArthur Foundation. (2013). *Towards the circular economy*. Retrieved from http://www.ellenmacarthurfoundation.org/business/reports/ce2013.

Escaler, M., & Teng, P. (2011). “Mind the gap”: Reducing waste and losses in the food supply chain food losses in the food supply chain. The Centre for Non-Traditional Security (NTS) Studies, S. Rajaratnam School of International Studies (RSIS), NTS Insight, 6. Retrieved from http://www.rsis.edu.sg/nts/HTML-Newsletter/Insight/pdf/NTS_Insight_jun_1101.pdf.

European Commission. (2010). Being wise with waste: The EU ’s approach to waste management (pp. 1–16). Retrieved from http://ec.europa.eu/environment/waste/pdf/WASTE%20BROCHURE.pdf.

European Commission Directorate-General for the Environment. (2000). *Success stories on composting and separate collection*, Brussels. Retrieved from http://ec.europa.eu/environment/waste/publications/pdf/compost_en.pdf.

European Parliament Resolution (2011/2175 (INI)). (2012). European Parliament resolution of 19 January 2012 on how to avoid food wastage: strategies for a more efficient food chain in the EU (2011/2175(INI)) (Vol. 2175(January), pp. 1–8). Retrieved from http://www.europarl.europa.eu/sides/getDoc.do?pubRef=-//EP//NONSGML+TA+P7-TA-2012-0014+0+DOC+PDF+V0//EN.

Eurostat. (2013a). *Children at risk of poverty or social exclusion*. Retrieved Dec 22, 2013, from http://epp.eurostat.ec.europa.eu/statistics_explained/index.php/Children_at_risk_of_poverty_or_social_exclusion.

Eurostat. (2013b). Environment in the EU27 (March, pp. 1–3). Retrieved from http://epp.eurostat.ec.europa.eu/cache/ITY_PUBLIC/8-04032013-BP/EN/8-04032013-BP-EN.PDF.

FareShare. (2013). *About us*. Retrieved Oct 24, 2013, from http://www.fareshare.org.uk/about-us-2/.

FEBA. (2013a). Food waste. Retrieved Oct 20, 2013, from http://www.eurofoodbank.eu/portail/index.php?option=com_content&view=article&id=146:halte-au-gaspillage-alimentaire&catid=27:lutte-contre-le-gaspillage&Itemid=46&lang=en.

FEBA. (2013b). *What do we do?* Retrieved Oct 20, 2013, from http://www.eurofoodbank.eu/portail/index.php?option=com_content&view=category&layout=blog&id=3&Itemid=8&lang=en.

Food Bargains. (2010). Food bargains—about us. Retrieved Apr 4, 2013, from http://www.foodbargains.co.uk/pages/food_bargains.html.

Food Recovery Committee. (2007). *Comprehensive guidelines for food recovery programs.* Retrieved from http://www.foodprotect.org/media/guide/food-recovery-final2007.pdf.

Foresight. (2010). *How can waste reduction help to healthily and sustainably feed a future global population of nine billion people?*, London. Retrieved from http://www.bis.gov.uk/assets/foresight/docs/food-and-farming/workshops/11-608-w4-expert-forum-reduction-of-food-waste.pdf.

Foresight. (2011). *The future of food and farming: Challenges and choices for global sustainability*, London. Retrieved from http://www.bis.gov.uk/assets/foresight/docs/food-and-farming/11-546-future-of-food-and-farming-report.pdf.

Gustavsson, J., et al. (2011). *Global food losses and food waste—extent, causes and prevention*, Rome. Retrieved from http://www.fao.org/docrep/014/mb060e/mb060e00.pdf.

Hodges, R. J., Buzby, J. C., & Bennett, B. (2011). Postharvest losses and waste in developed and less developed countries: Opportunities to improve resource use. *Journal of Agricultural Science*, *149*, 37–45. Retrieved from http://journals.cambridge.org/download.php?file=%2FAGS%2FAGS149_S1%2FS0021859610000936a.pdf&code=16a2afa2796c12d2e4a6c55bf429b91c.

Institution of Mechanical Engineers. (2013). *Global food waste not, want not*, London. Retrieved from http://www.imeche.org/Libraries/Reports/Global_Food_Report.sflb.ashx.

Kantor, L.S. et al.. (1997). Estimating and Addressing America's Food Losses. *Food Review, 1264* (202), 2–12. Retrieved from http://www.calrecycle.ca.gov/reducewaste/food/foodlosses.pdf.

Karim Ghani, W. A. W. A., et al. (2013). An application of the theory of planned behaviour to study the influencing factors of participation in source separation of food waste. *Waste management (New York, N.Y.)*, *33*(5), 1276–1281. Retrieved Aug 16, 2013, from http://www.ncbi.nlm.nih.gov/pubmed/23415709.

Lamb, G. & Fountain, L. (2010). *An investigation into food waste management.* Retrieved from http://www.actiondechets.fr/upload/medias/group_b_report_compressed.pdf.

Lin, C. S. K. et al. (2013). Food waste as a valuable resource for the production of chemicals, materials and fuels. Current situation and global perspective. *Energy & Environmental Science*, *6*(2), 426–464. Retrieved Aug 13, 2013, from http://xlink.rsc.org/?DOI=c2ee23440h.

Lipinski, B., et al. (2013). *Reducing food loss and waste*, Washington, DC. Retrieved from http://www.worldresourcesreport.org.

Maxwell, D., et al. (2011). *Addressing the rebound effect.* Retrieved from http://ec.europa.eu/environment/eussd/pdf/rebound_effect_report.pdf.

Nellemann, C., et al. (2009). *The environmental food crisis—the environment's role in averting future food crises.* Retrieved from http://www.grida.no/files/publications/FoodCrisis_lores.pdf.

Parfitt, J., & Barthel, M. (2011). *Global food waste reduction: Priorities for a world in transition*, London. Retrieved from ssets/foresight/docs/food-and-farming/science/11-588-sr56-global-food-waste-reduction-priorities.pdf.

Parfitt, J., Barthel, M., & Macnaughton, S. (2010). Food waste within food supply chains: quantification and potential for change to 2050. *Philosophical Transactions of the Royal Society of London. Series B, Biological sciences*, *365*(1554), 3065–3081. Retrieved Feb 28, 2013, from http://www.pubmedcentral.nih.gov/articlerender.fcgi?artid=2935112&tool=pmcentrez&rendertype=abstract.

Pfaltzgra, L. A., et al. (2013). Food waste biomass: a resource for high-value chemicals. *Green Chemistry*, *15*, 307–314. Retrieved from http://pubs.rsc.org/En/content/articlepdf/2013/gc/c2gc36978h.

Redlingshöfer, B., & Soyeux, A. (2012). Food losses and wastage as a sustainability indicator of food and farming systems. Retrieved from http://www.ifsa2012.dk/downloads/WS6_1/Redlingshofer.pdf.

Schneider, F. (2008). *Wasting food—an insistent behaviour*, Alberta, Canada. Retrieved from http://www.ifr.ac.uk/waste/Reports/Wasting%20Food%20-%20An%20Insistent.pdf.

Sonigo, P., et al. (2012). *Assessment of resource efficiency in the food cycle, final report*, Retrieved from http://ec.europa.eu/environment/eussd/pdf/foodcycle_Final%20report_Dec%202012.pdf.

The Joint Food Wastage Declaration "Every Crumb Counts." (2013). *The joint food wastage declaration "every crumb counts."* (pp. 1–8). Retrieved from http://everycrumbcounts.eu/uploads/static_pages_documents/JD_PDF_%28FINAL_VERSION%291.pdf.

UK Department of Energy and Climate Change & Defra. (2011). *Anaerobic digestion strategy and action plan*, London, UK. Retrieved from https://www.gov.uk/government/uploads/system/uploads/attachment_data/file/69400/anaerobic-digestion-strat-action-plan.pdf.

UK Government. (2010). *Food 2030*. Retrieved from http://archive.defra.gov.uk/foodfarm/food/pdf/food2030strategy.pdf.

U.S. EPA. (2006). *Food scraps go to the animals Barthold recycling and roll-off services*. Retrieved from http://www.epa.gov/foodrecovery/success/barthold.pdf.

U.S. EPA. (2012). Composting. *Agriculture*. Retrieved Nov 5, 2013, from http://www.epa.gov/agriculture/tcop.html.

U.S. EPA. (2013a). Feed animals. *Wastes—resource conservation—food waste*. Retrieved Nov 2, 2013, from http://www.epa.gov/foodrecovery/fd-animals.htm.

U.S. EPA. (2013b). *Municipal solid waste generation, recycling, and disposal in the United States: Facts and figures for 2011*, Washington, DC. Retrieved from http://www.epa.gov/osw/nonhaz/municipal/pubs/MSWcharacterization_508_053113_fs.pdf.

U.S. EPA. (2013c). Types of composting. *Wastes—resource conservation*. Retrieved Nov 5, 2013, from http://www.epa.gov/wastes/conserve/composting/types.htm.

Ujhelyi, K. (2013). *Forward survey report*. Retrieved from http://foodrecoveryproject.eu/wp-content/uploads/2012/11/FoRWaRd-D3.3_Report_of_Analysis_of_Results.pdf.

Value Chain Management Centre. (2012). *Cut waste, grow profit*. Retrieved from http://www.valuechains.ca/usercontent/documents/Cut%20Waste%20Grow%20Profit%20FINAL%20DOCUMENT%20Oct%203%2012.pdf.

Viel, D. (2011). *Food wastage study mid-term report*. Retrieved from http://www.developpement-durable.gouv.fr/IMG/food%20waste%20mid-term%20report_VF.pdf.

Waarts, Y., et al. (2011). *Reducing food waste; obstacles experienced in legislation and regulations*, Wageningen. Retrieved from http://edepot.wur.nl/188798.

Weber, B., Herrlein, S., & Hodge, G. (2011). *The challenge offood waste*, London. Retrieved from www.planetretail.net.

Wells, P., Stone, I., & Coss, D. (2011). *Evaluation offood waste collections*. Retrieved from http://www.wlga.gov.uk/archive-of-reports9/wlga-report-evaluation-of-food-waste-collectio.

WRAP. (2009). *Food waste collection guidance*. Retrieved from http://www.wrap.org.uk/sites/files/wrap/food%20waste%20collection%20guidance%20-%20amended%20Mar%202010_1.pdf.

WRAP. (2011). *Investigation into the possible impact of promotions on food waste*. Retrieved from http://www.wrap.org.uk/sites/files/wrap/WRAP%20promotions%20report%20FINAL%20241111.pdf.

WRAP. (2012). *Decoupling of waste and economic indicators*. Retrieved from http://www.wrap.org.uk/sites/files/wrap/Decoupling%20of%20Waste%20and%20Economic%20Indicators.pdf.

Chapter 5
Research Methods

The study entails three major components. The first component is focused on the investigation of sources in regard with the state-of-the-art of the food waste problem at the global level, available and applied methods of its reduction, related case studies and best practices, as well as on theoretical base for the development of the food waste management tools.

The second component seeks to identify the problem of food waste in seven counties of the Baltic Region. It centres on the analysis of data of the economic situation in each country, particularly consumer purchasing power, undernourishment and poverty level, food consumption pattern, as well as, on the current situation regarding biodegradable waste management and renewable energy production. The research on the issue of food waste covers such subjects as the state of the problem in each country, its causes, existing food waste prevention initiatives, including awareness campaigns, industrial uses, redistribution programmes, informational tools, training programmes, logistical improvements, as well as, existing obstacles, related legislation and required further actions to tackle the problem.

The information and data collection process comprised the following stages:

The first stage was the search that covered documents (e.g. reports, studies, articles, legislative documents: regulations, directives, normative acts) and data, available in the public domain and in academic literature. It targeted websites of organisations associated with biodegradable waste, food waste and food waste projects at the global and national levels (e.g. FAO, World Bank, UN, FEBA, U.S. EPA, EEA, European Commission, Food Banks and etc.) as well using internet search engines, Google Scholar and proprietary academic publication databases such as Science Direct. The statistical data were obtained from statistical databases such as Eurostat, FAOSTAT, World Bank statistical database, databases of the national statistical offices of each country.

It should be noted that the search covered sources available in English and Russian languages.

The next stage was an attempt to refine the aforementioned search by examining the references cited in the found documents. The bibliographies, and any footnote

W. Leal Filho and M. Kovaleva, *Food Waste and Sustainable Food Waste Management in the Baltic Sea Region*, Environmental Science and Engineering, DOI 10.1007/978-3-319-10906-0_5

references, in documents collected during the first stage were used to expand the pool of potentially useful data sources.

Another stage included interviews with representatives of organisations that are active in the field of food waste in each country, e.g. Food Banks. The interviews were conducted via e-mail. The interviewees were asked about the state of the problem of food in their countries, and also to provide statistical data, if it was not against the confidentiality policy of their organisation. An additional component was a survey. For this purpose, a questionnaire in English and Russian was developed. The questionnaire was designed to get more detailed information about the degree to which the problem is discussed in public, the involvement of the government, and possible causes of food waste in a country. Examples of the questionnaire are included in Appendices A and B.

To see the activities of a food bank in practice, two distribution points of one of the German Food Banks were visited.

It should also be noted that while the methodology is presented in a linear format, the process was iterative and some tasks occurred in parallel. New data sources and candidates for interviews were continually identified as the research progressed.

To define the scope of the current study and to set its boundaries, the topology of the biodegradable waste was developed. The main purpose of such hierarchy is to clearly indicate the place of food waste in the biodegradable waste classification, as well as to define food waste sub-categories.

The calculation of food waste fractions was made based on the waste statistical data, provided in the FAO food balance sheets and on the FAO technical conversion factors such as extraction rates.

The fraction of food waste, calculated based on the FAO food balance sheets, is the difference between a minimum amount of waste generated per one hundred tonnes among the discussed countries and an amount of the same type of waste generated in each of the rest of the countries. The minimum amount is assumed to be an amount of unavoidable waste.

The fraction of food waste, calculated based on the FAO extraction rates, is the difference between the maximum and actual values of the extraction rates existing in each country. The maximum value is assumed to be the maximum possible extraction rate, limited, only by currently existing technologies in one of the countries.

Microsoft Excel was used as a main software application to make calculations and build charts.

The third component of the study includes the analysis of available information, statistical data, best practice and existing treatment methods that has become a base for development of recommendation on steps, required to be taken in the near future order to change the food waste situation in each country.

The choice methods used in this study had taken into account that a lack of willingness to provide data/information from many persons could pose a problem, which was compensated by a combination of the use of primary and secondary data.

Chapter 6
Overview of the Baltic Region Countries

The current work is focused on seven countries of the Baltic region, such as Belarus, Estonia, Germany, Latvia, Lithuania, Poland and Sweden. Despite a number of differences in population size, land area, density (Table 6.1) and GDP per capita (Fig. 6.1), in the area of waste management the countries have a number of similarities, for example, the same waste legislation, except Belarus, and similar historical background, excepting Germany and Sweden, countries which, in way or another, have influenced and shaped society's attitude towards the food and waste issues as well as consumer behaviour. All these make a comparative analysis of the following countries regarding the problem of the food waste management and its possible future trends interesting and useful.

6.1 Main Economic Activities

6.1.1 Belarus

The main industries in the country are of the following types:

- Metallurgical;
- mechanical engineering, including tractors and agricultural, cars, machine-tool constructing and tool industry, instrument making, radio engineering, electro technical, electronic, optics-mechanical industry;
- metal working;
- chemical and petrochemical;
- light industry;
- food industry;

The agricultural sector is led by grain, potatoes, vegetables, sugar beet, flax, meat and dairy industries.

In 2012, among the main export articles were mineral products, chemical industry production, rubber, cars, equipment and vehicles (Fig. 6.2) that constituted more than 70 % of the total export of the country (Official Website of the Republic of Belarus 2013).

W. Leal Filho and M. Kovaleva, *Food Waste and Sustainable Food Waste Management in the Baltic Sea Region*, Environmental Science and Engineering, DOI 10.1007/978-3-319-10906-0_6

Table 6.1 General information about the countries (Eurostat 2013a, b)

Indicators	Belarus	Estonia	Germany	Latvia	Lithuania	Poland	Sweden
Land Area (sq. km)[a]	202,910	42,390	348,570	62,200	62,674	304,150	410,340
Population (milllion), 2013	9.464[b]	1.325	80.524	2.024	2.972	38.533	9.556
Density (inhabitants per sq. km), 2011	46.7[c]	30.9	229	33.1	48.3	123.2	23
GDP (billio US$), 2012[c]	63.27	21.85	3399.59	28.32	42.09	489.80	525.74
% of rural population[d]	25.0	30.5	26.1	32.3	32.9	39.1	14.8
% of urban population[d]	75.0	69.5	73.9	67.7	67.1	60.9	85.2

[a] Land area is a country's total area, excluding area under inland water bodies, national claims to continental shelf, and exclusive economic zones. In most cases the definition of inland water bodies includes major rivers and lakes. *Source* World Bank (2013a, b)

[b] For the year 2012, *Source* World Bank (2013a, b)

[c] *Source* World Development Indicators

[d] For the year 2011, *Source* (FAO 2013a)

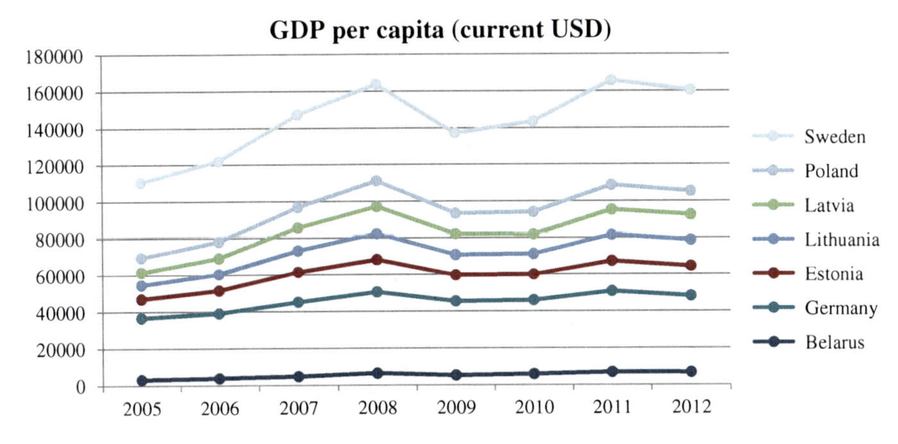

Fig. 6.1 GDP per capita in the selected countries, current USD (World Bank 2013a, b)

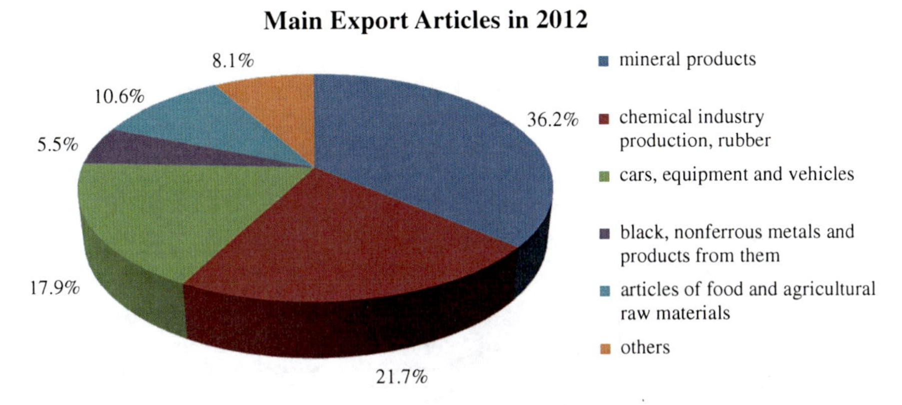

Fig. 6.2 Commodity structure of export in 2012 (Official Website of the Republic of Belarus 2013)

6.1.2 Estonia

The important sectors of the Estonian economy are the processing industry, transport, warehousing and communications, rental and letting, commerce and estate, and business services (Lauri 2012). 17 % of the processing industry belongs to the local food industry that focuses on the dairy, meat, bakery and fish products (Fig. 6.3) (Estonian Ministry of Agriculture 2010).

The country exports agricultural and industrial machinery, telecommunication, construction engineering and electrical equipment, as well as food, beverages and tobacco (Bank of Lithuania 2013). The volume of export of goods and services

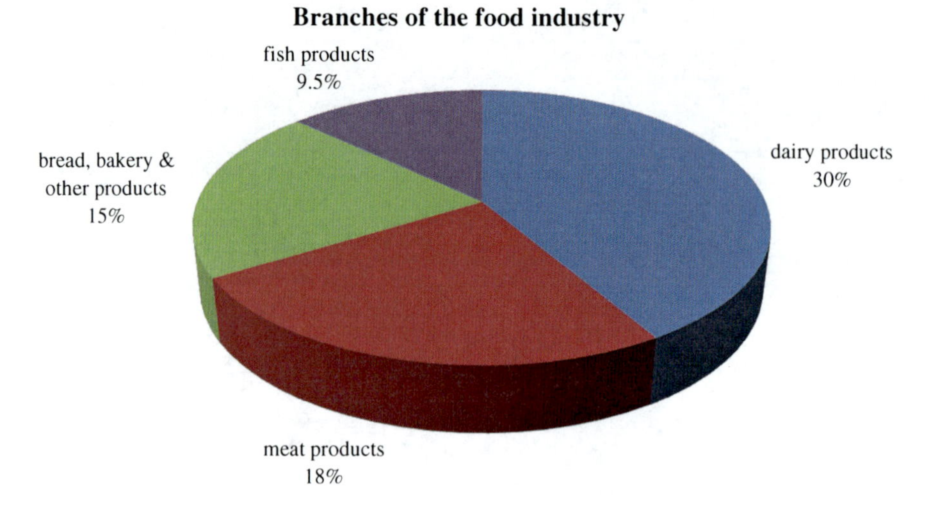

Fig. 6.3 Branches of food industry (Estonian Ministry of Agriculture 2010)

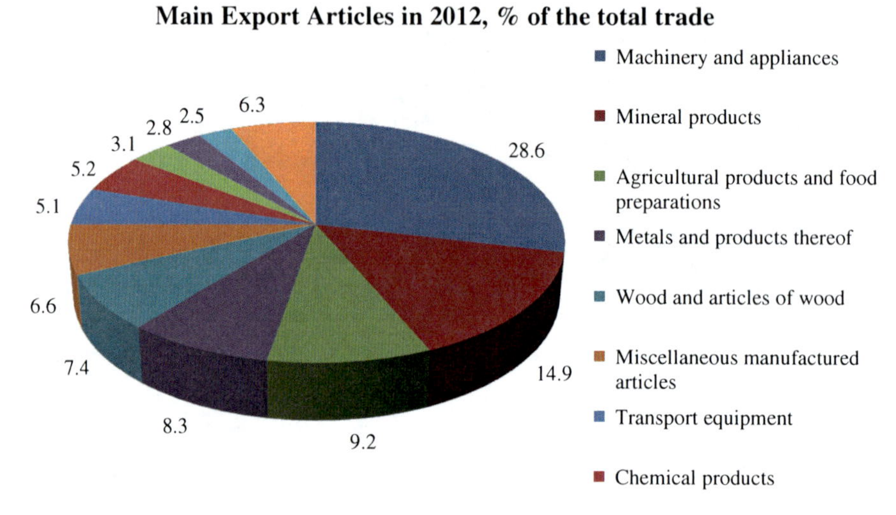

Fig. 6.4 Main Export Articles in 2012, as a percentage of the total trade (Estonia.eu 2013)

amounts to 73 % of the Estonian GDP (Lauri 2012). According to the Statistical Office of Estonia, in 2012, among the main export articles were machinery and appliances, and mineral products with 28.6 and 14.9 % of the total trade, respectively (Fig. 6.4).

6.1.3 Germany

As well as having the largest population among the EU countries (Table 6.1), Germany is the world's third largest economy, producing among others products automobiles, precision engineering products, electronic and communications equipment, chemicals and pharmaceuticals (European Union 2013). Its major industries with the highest gross value added in 2012 were (German Federal Statistical Office 2013b)

- manufacturing
- public services, education, health
- trade, transport, accommodation and food services
- real estate activities

According to the Federal Statistical Office, in 2012, among the main exports articles were motor vehicles (17.4 %), machinery and equipment (15.1 %), chemicals and chemical products (9.6 %), computer, electronic and optical products (7.9 %), and food products (4.2 %) (German Federal Statistical Office 2013a).

6.1.4 Latvia

The main sectors of the Latvian economy are food and beverage production, production of wood products, production of pharmaceuticals, chemicals and chemical products, production of metals and metal products, production of computers, electronic and optical equipment, telecommunications, software development and ancillary services for transport and logistics (BalticExport 2013). According to the Latvian Central Statistical Bureau, in 2011, wood, paper, base metals and machinery, and electric equipment comprised the main export sectors (Fig. 6.5).

The food industry, the second largest sector of the Latvian economy, focuses on meat, milk and dairy production (Vorne et al. 2011).

6.1.5 Lithuania

The largest and most populous of the three post-soviet Baltic countries (Tylaite and Bastys 2013), Lithuania exhibits the strongest growth in household consumption in comparison to Latvia and Estonia (Rudzitis et al. 2013).

The main sectors of the national economy are (London Chamber of Commerce and Industry 2010)

- agriculture: cereals, grain, fodder and rape seed are the most commonly grown agricultural product;

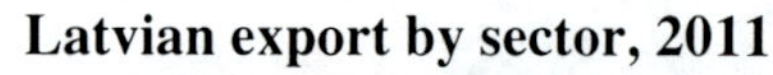

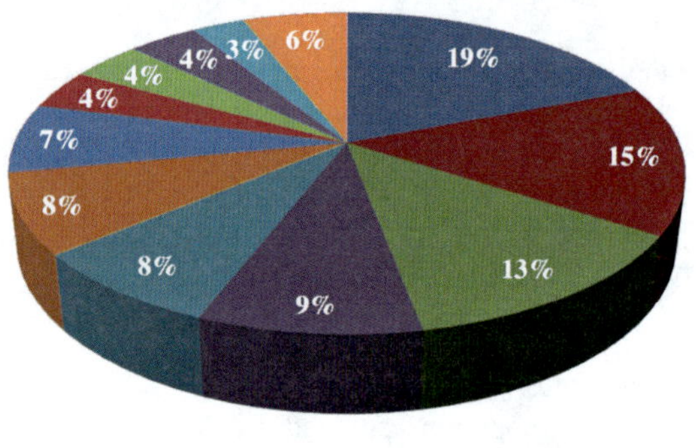

Fig. 6.5 Main export articles in 2011 (Ministry of Economics of the Republic of Latvia 2013)

- energy industry: it is the only post-soviet Baltic state with a nuclear power station;
- laser technology industry: the country has established itself as a global leader in the laser technology industry;
- manufacturing industry: various sub-sectors such as the manufacturing of furniture, machinery and electronic equipment and the manufacturing of plastics;
- services and ICT industry;
- transport industry

The country mainly exports food, beverages and tobacco, chemical, rubber, plastics and non-metallic minerals (Bank of Lithuania 2013).

6.1.6 Poland

Rich in natural mineral resources, including iron, zinc, copper and rock salt (European Union 2013), the country manufactures metal products, furniture, pharmaceutical products, computers, electronic and optical goods, machinery and equipment. The dominating commodity groups of Polish export are electric machinery and chemical products (Ministry of Economy of the Republic of Poland 2012). According to the Polish Central Statistical office, in 2013, the main export articles were machinery and transport equipment (37.46 %), and manufactured goods (20.69 %) (Fig. 6.6).

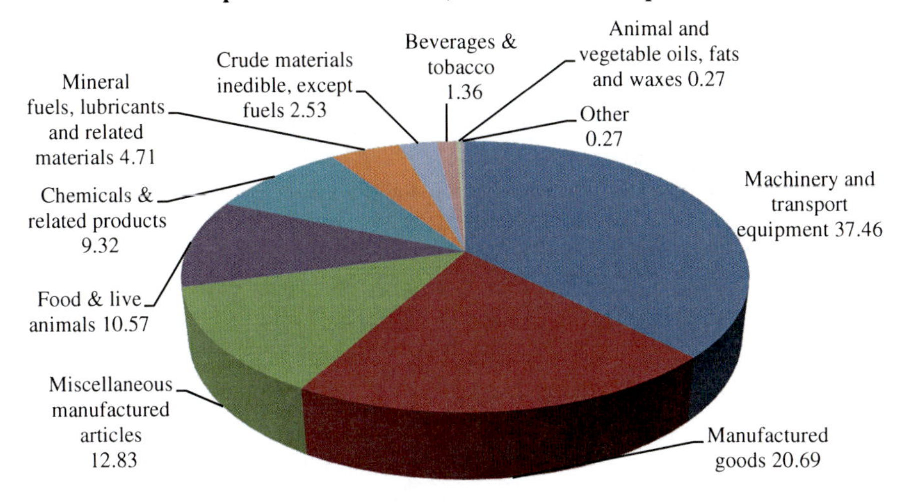

Fig. 6.6 Main export articles in 2013, percentage share (Central Statistical Office of Poland 2013)

6.1.7 Sweden

Sweden has the largest population of the Nordic countries. The southern part of the country is chiefly agricultural (European Union 2013). Among dominating sectors of the national economy are manufacturing, business and financial services, wholesale and retail industries (Fig. 6.7) (McKinsey and Company 2012). The country exports cars, engineering products, steel, electronic devices, communications equipment and paper products (European Union 2013).

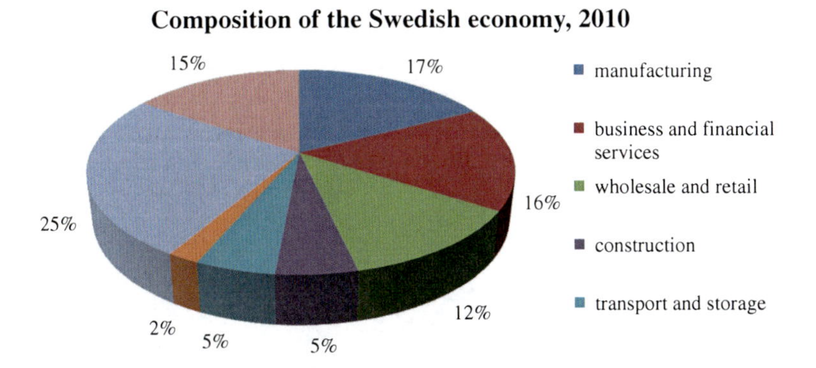

Fig. 6.7 Composition of the Swedish economy according to the gross value added in 2010 (McKinsey & Company 2012)

6.2 Renewable Energy

The Renewable Energy Directive (Directive 2009/28/EC) sets legally binding renewable energy targets for the Member States to get 20 % of their energy production from renewable sources by 2020. The sources include wind, solar, hydroelectric and tidal power as well as geothermal energy and biomass. According to the latest Eurostat data, in 2011, Latvia, Lithuania and Sweden had the biggest share of primary production of renewable energy of total primary energy production with 99.8, 90.1, 49.1 %, respectively (Fig. 6.8).

In the same year, Estonia reached 96.5 % of renewable energy primary production from biomass and waste, followed by Poland with 93.3 % and Lithuania with 92.7 % (Fig. 6.9).

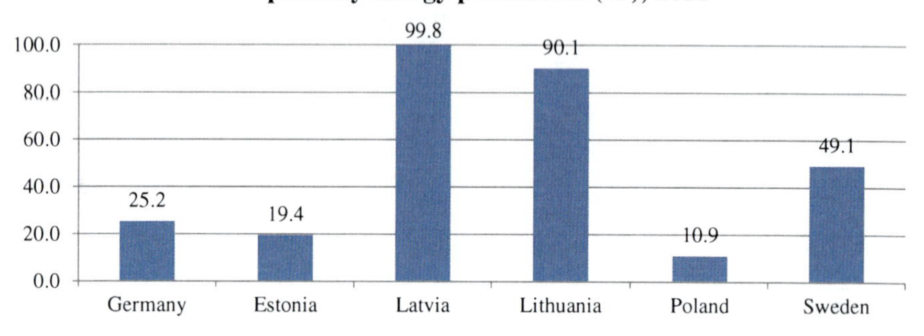

Fig. 6.8 Share of primary production of renewable energy of total primary energy production in the selected countries in 2011 (Eurostat 2013a, b)

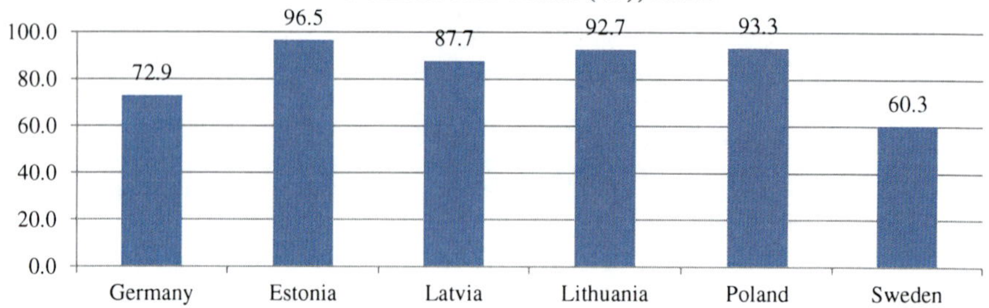

Fig. 6.9 Share of renewable energy primary production from biomass and waste in 2011 (Eurostat 2013a, b)

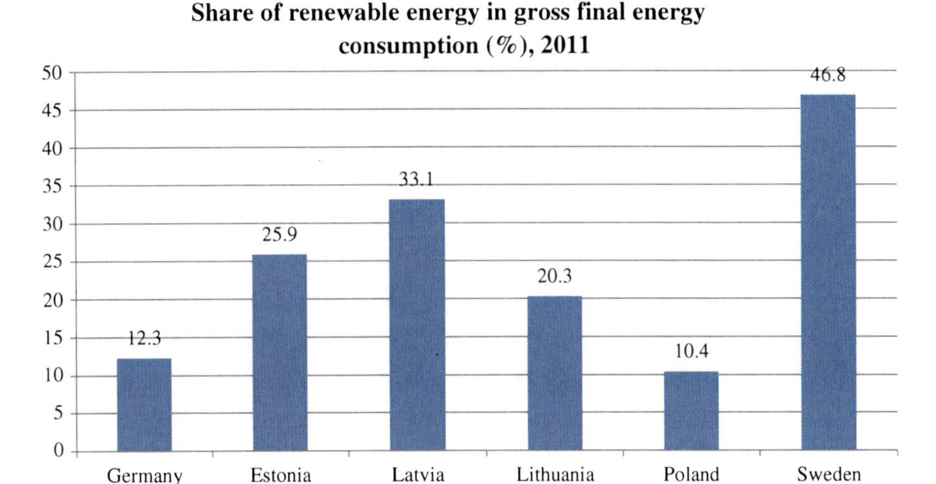

Fig. 6.10 Share of renewable energy in gross final energy consumption in 2011 (Eurostat 2013a, b)

In 2011, Sweden, Latvia and Estonia had the largest share of renewable energy in the gross final energy consumption 46.8, 33.1 and 25.9 %, respectively (Fig. 6.10).

The situation in Belarus is slightly different. The country depends on Russian crude oil and natural gas which are imported at prices substantially below the world market (Index Mundi 2013). In 2010, Belarus started the production of biogas. According to the National Statistical Committee, in 2012, the amount of biogas production is doubled in comparison to 2010. In addition, there was a sharp increase in usage of hydropower and wind energy (National Statistical Committee of the Republic of Belarus 2013a).

6.3 Food Consumption and Undernourishment

The largest dissimilarities among the discussed countries are in the food consumption pattern that to a large extent is influenced by poverty, undernourishment levels and a share of consumer expenditures spent on food.

6.3.1 Poverty Level

According to the Eurostat data, in 2012, in Latvia, the percentage of people found at risk of poverty or social exclusion was twice bigger than in Sweden (Fig. 6.11). In Poland 2.5 million people live in deep poverty and more than 6 million struggle

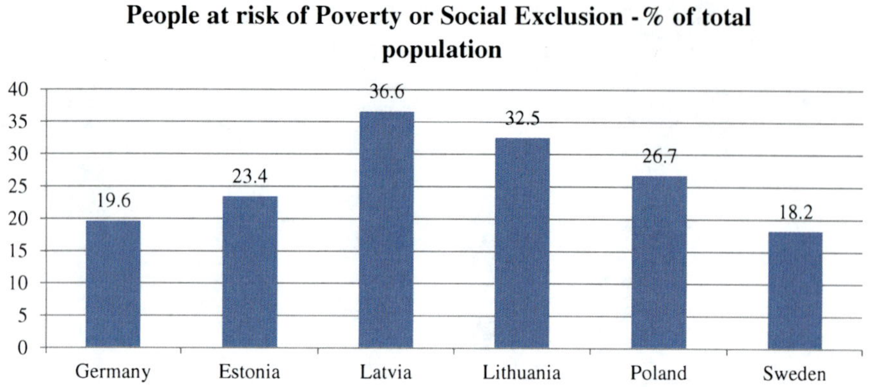

People at risk of Poverty or Social Exclusion - % of total population

	Germany	Estonia	Latvia	Lithuania	Poland	Sweden
%	19.6	23.4	36.6	32.5	26.7	18.2

Fig. 6.11 People at risk of poverty or social exclusion—percentage of total population, 2012 (Eurostat 2013a, b)

financial difficulties and in many cases need food support (Gosiewska 2013). The Estonian statistical office indicates that in 2012, 18.7 % of the Estonian population lived in relative poverty and 7.3 % in absolute poverty (Statistics Estonia 2013).

In Belarus, the results of the study conducted by the Research Center of the Institute for Privatization and Management showed that in 2012, the levels of absolute and relative poverty reached 6.6 % and 11.2 % of total population (Research Center of the Institute for Privatization and Management 2013).

6.3.2 Undernourishment

According to the World Bank, in 2011, an average 5 % of total population in the discussed countries were undernourished (World Bank 2013b). In 2012, the highest values of depth of the food deficit, expressed in kilocalories per person per day, were in Latvia (28) and Estonia (23) (Fig. 6.12).

Furthermore, in 2009, both countries had the largest percentage of children (1–15 years), 15.4 % and 10.4 %, who did not eat fresh fruit and vegetables once a day as these items could not be afforded, in comparison to 2.5 % in Germany. Similarly, in Latvia 10.6 % and Lithuania 9 % of children did not eat one meal with meat, chicken or fish, or vegetarian equivalent (proteins) per day because a household could not afford it (Fig. 6.13).

In Belarus, according to the Research Center of the Institute for Privatization and Management, in 2012, 66.3 thousand people (about 0.7 % of total population in 2011) got an food allowance for children under the age of 2 years (Research Center of the Institute for Privatization and Management 2013).

It is also important to note that one of the integral parts of the national food production are personal subsidiary plots. In 2012, 91.3 % of total population got an income in the physical terms (animal, dairy or/and vegetable products) from their or

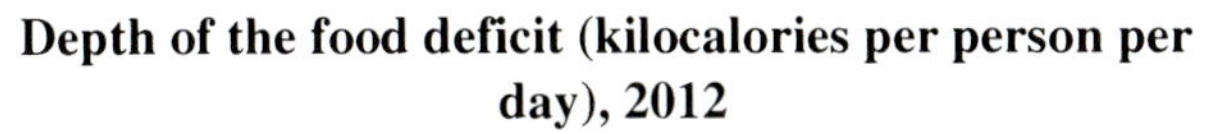

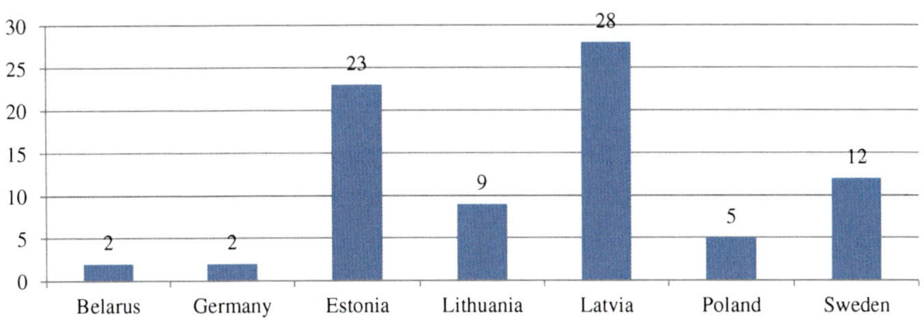

Fig. 6.12 Depth of the food deficit (kilocalories per person per day) in 2012 (World Bank 2013a)

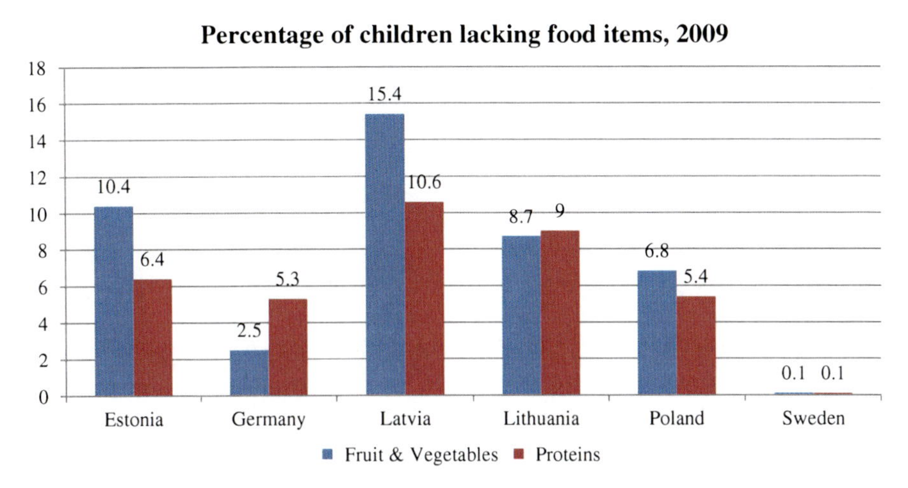

Fig. 6.13 Percentage of children (1–15 years) deprived (lacking food items), in 2009 (Eurostat 2013a) * For Sweden data are unreliable

their relatives subsidiary plots, in 2011, this number reached 93.3 % (Research Center of the Institute for Privatization and Management 2013).

6.3.3 European Food Aid

In the frame of the common agricultural policy (CAP), all countries, except Sweden and Germany, get assistance as a part of the CAP Food programme. In 2011, the 'leader' among the discussed countries was Poland. The country got 14.5 % out of the total CAP expenditures for this programme in the EU (Fig. 6.14).

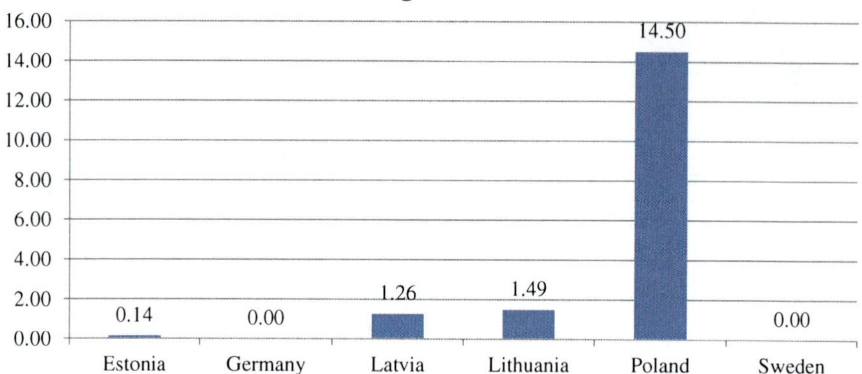

Fig. 6.14 Percentage of total CAP expenditures on Food Programs in 2011 in EU (European Commission 2013a)

6.3.4 Food Expenditures

The share of consumer expenditure spent on food varies greatly among the countries. According to the Economic Research Service (ERS) of the US Department of Agriculture (USDA), in Belarus, consumer expenditures on food consumed at home amount to 36.1 % of total expenditures, whereas in Sweden and Germany consumers spend only 12.2 % and 10.9 % respectively (Fig. 6.15).

Despite the Eurostat data on a percentage share of household expenditures on food out of total final consumption expenditures slightly differ, both sources indicate that consumers in Germany and Sweden spend least (Fig. 6.16).

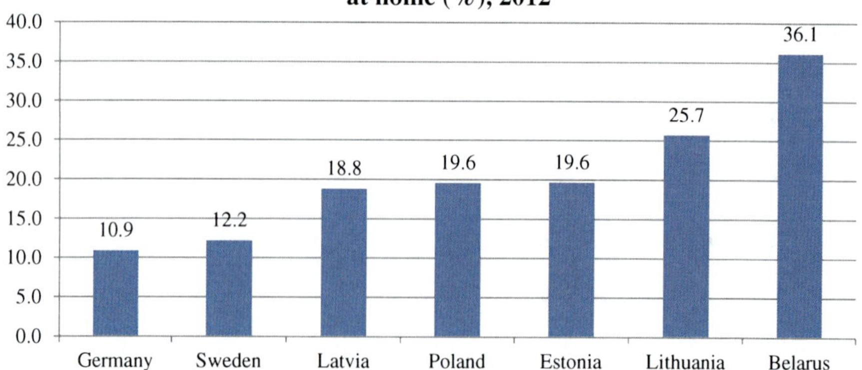

Fig. 6.15 The percentage share of consumer expenditures spent on food consumed at home in the selected countries in 2012 (ERS/USDA 2013)

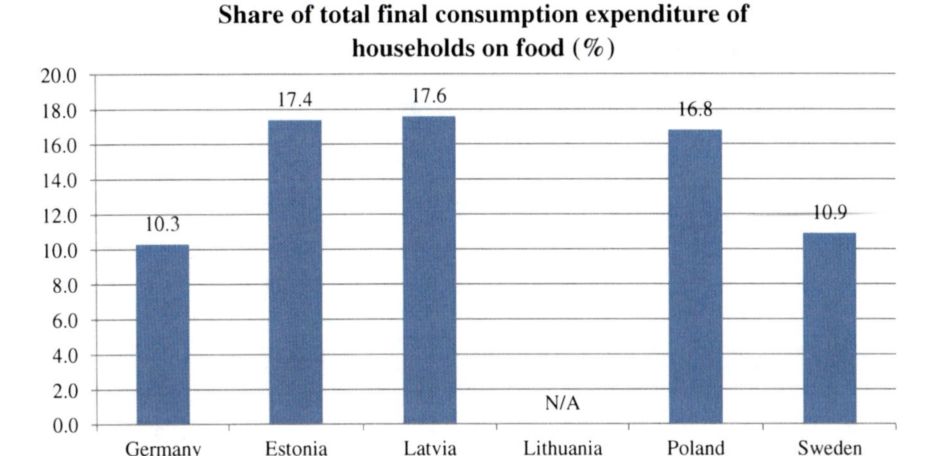

Fig. 6.16 Percentage share of the total final consumption expenditure of households on food (Eurostat 2013a, b)

In the case of Latvia, Lithuania and Estonia, the data regarding household expenditures on food vary among different sources. According to the study conducted by the Institute for Private Finances, Swedbank, in 2012, the largest proportion of expenditures (almost 26 %) on food and non-alcoholic beverages is in Lithuania, and the smallest (almost 21 %) in Estonia. Latvia's expenditures on food and non-alcoholic beverages amount to almost 24 % of total expenditures (Institute for Private Finances, Swedbank 2013). In 2013, the values have almost not changed, the weight of food items in a consumer basket comprised 25.3, 23.5 and 21.3 % in Lithuania, Latvia and Estonia respectively (Swedbank Macro Research 2013). It is also important to note that in Estonia 78 % of consumers buy their food mostly from big food markets (Esko et al. 2012).

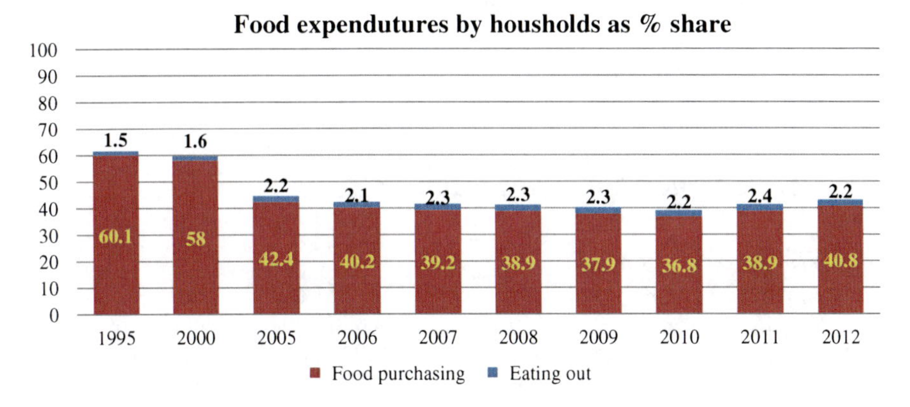

Fig. 6.17 Food expenditures by households as a percentage share (National Statistical Committee of the Republic of Belarus 2013b)

The structure of the household expenditures in Belarus is very different. Despite the significant decrease of food expenditures by almost 20 % in comparison to 2000, in 2012, a consumer in Belarus spent almost four times more than a consumer in Germany or Sweden (Fig. 6.17).

6.4 Biodegradable Waste

6.4.1 Legislation

Although all of the countries in this study apart from Belarus are obliged to follow the same waste legislation, differences in the initial situation, treatment methods and available facilities define various ways to achieve the targets set by the EU waste legislation. Following the requirements of the Landfill Directive in regards with the allowed amount of municipal biodegradable waste to be landfilled, Germany has met the 2016 target in 2006 (Fischer 2013b), whereas Estonia has met the 2013 target in 2009 (Fischer 2013a). Furthermore, Estonia has set national targets stricter than those in the Directive. The limits for landfilled biodegradable municipal waste are the following:(Fischer 2013a)

- 45 % by weight of total landfilled MSW from 2010;
- 30 % by weight of total landfilled MSW from 2013;
- 20 % by weight of total landfilled MSW from 2020

In Germany there are strict quality requirements for composts and digestates with regard to their impurity and contaminant content. Home composting is encouraged, but also regulated (Kern et al. 2012).

Starting from 2002 Latvia (BiPRO 2012b) and from 2003 Lithuania (ECN 2013) have banned landfilling of biodegradable waste. In Poland the ban on landfilling of separately collected biodegradable waste is in effect from January 2013 (Guziana et al. 2012). Sweden banned the landfill of sorted combustible waste in 2002 and of organic waste in 2005. The country has already reached all diversion targets of the Landfill Directive (Milios 2013).

6.4.2 Waste Generation and Treatment

Total Waste Generation

An amount of waste generated in a country is affected by a variety of factors, especially, by the state of a national economy. Among the discussed countries, in 2010, the biggest amount of waste was generated by Germany (approx. 364 million tonnes) and the smallest by Latvia (approx.1.5 million tonnes) (Fig. 6.18).

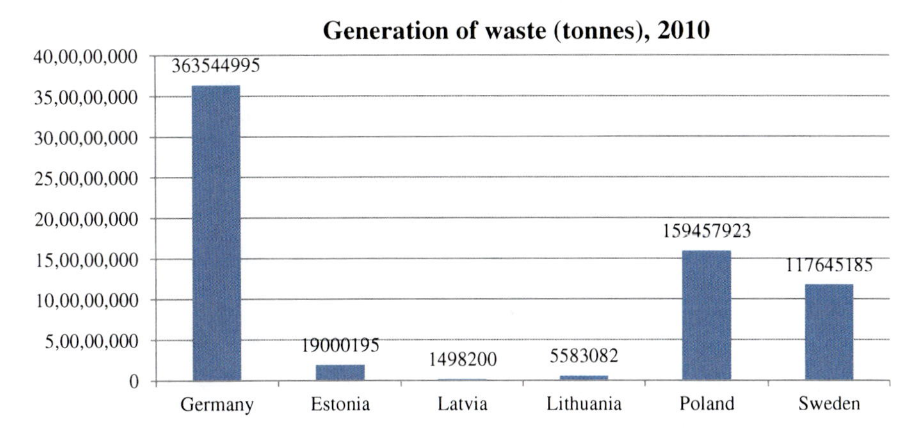

Fig. 6.18 Total amount of waste generated by households and businesses by economic activity, including hazardous and non-hazardous waste in 2010 (Eurostat 2013a, b)

In 2011, Germany also generated the biggest amount of MSW per capita, followed by Sweden and Lithuania (Fig. 6.19).

According to the Ministry of Natural Resources and Environmental Protection of the Republic of Belarus and the Scientific-Research Economic Institute of the Ministry of Economy of Belarus, during the years 2005–2010, the generated amount of MSW had steady growth and reached more than 3.7 million tonnes in 2010 (Fig. 6.20).

Animal and Vegetal Wastes

The situation with regard to the generation of animal and vegetal wastes is slightly different. According to the Eurostat, in 2010, Latvia and Lithuania had the highest share of animal and vegetal waste generated in relation to the total amount

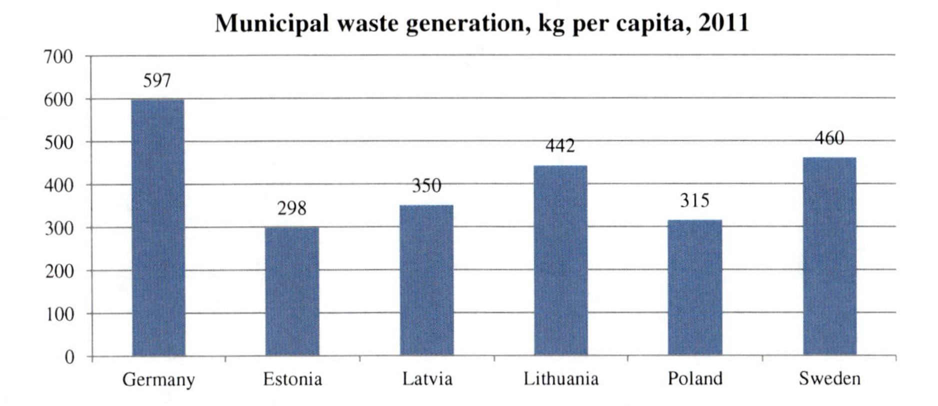

Fig. 6.19 Municipal waste generation, kg per capita in 2011 (Eurostat 2013a, b)

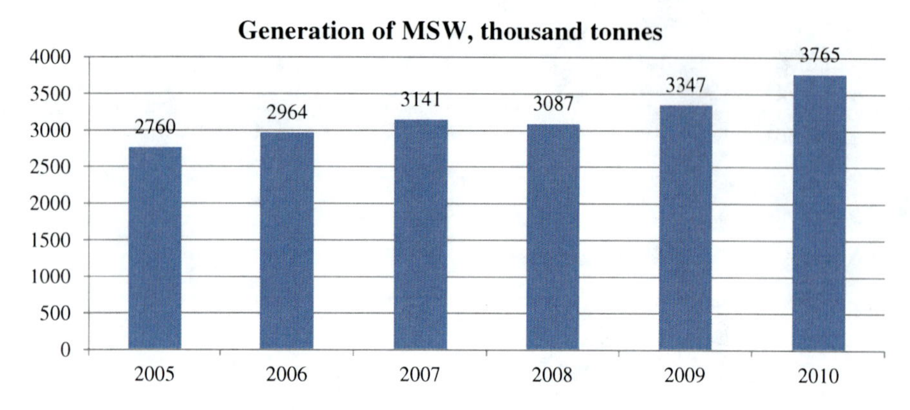

Fig. 6.20 Generation of MSW in Belarus (Ministry of Natural Resources and Environmental Protection of the Republic of Belarus 2010; Scientific-Research Economic Institute of the Ministry of Economy of Belarus 2012)

of waste, 11.1 and 9.59 % respectively. However, in absolute terms, Germany and Poland generated the biggest amounts of this type of waste (Fig. 6.21).

Total waste treatment

Despite a relative variety in waste treatment methods, according to the statistical data, the countries choose either disposal or recovery other than energy recovery methods (Fig. 6.22).

In comparison with the rest of the countries the percentage of MSW being composted and digested in 2011 stays very low, particularly in Latvia and Lithuania (Fig. 6.23).

Additional important aspects regarding the generation and treatment of biodegradable waste in the discussed countries are the following:

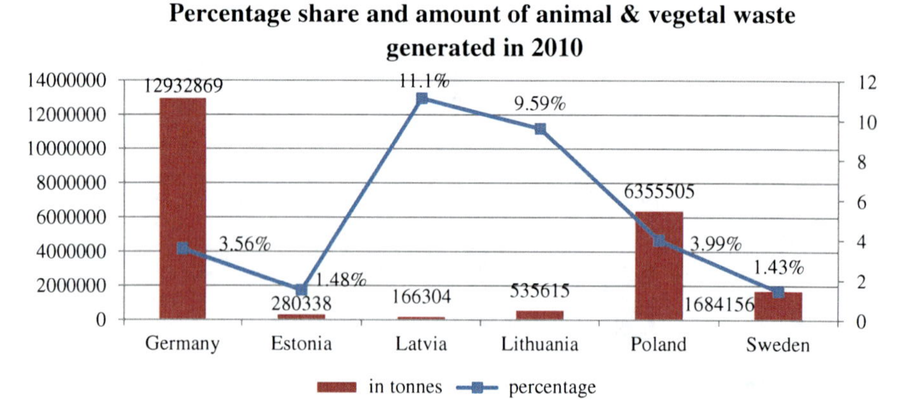

Fig. 6.21 Percentage share of animal and vegetal waste generated out of the total amount of waste and amount of animal and vegetal waste in tonnes generated in 2010 (Eurostat 2013a, b)

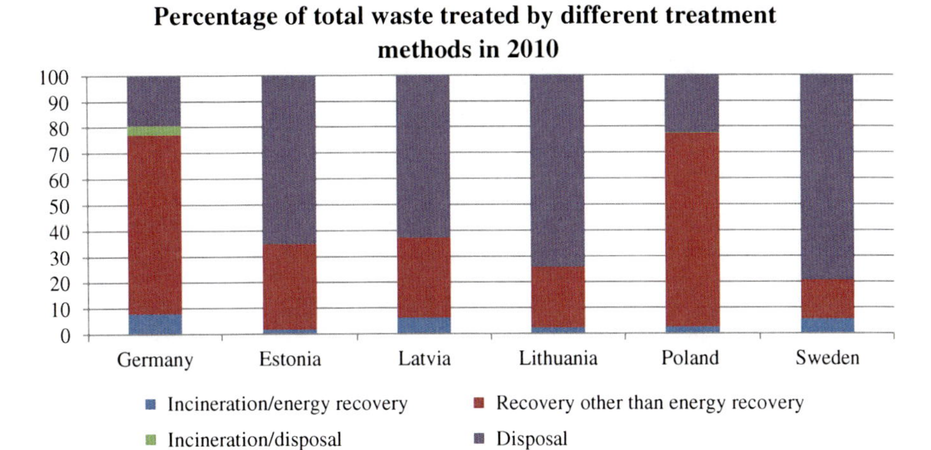

Fig. 6.22 Percentage of total waste treated by different treatment methods in 2010 (Eurostat 2013a, b)

6.4.2.1 Belarus

Currently in the country are operating

- 5 waste treatment plants;
- 7 biogas systems, including two landfill gas power plants;
- 3 biogas complexes which use agricultural residues;
- more than 50 production facilities at forest enterprises for processing low value raw wood material and timber waste for production of wood chips, wood briquettes and pellets

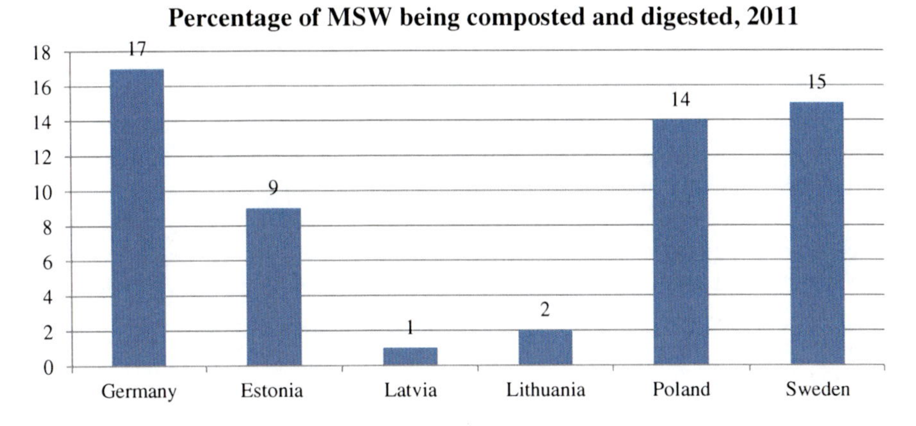

Fig. 6.23 Percentage of MSW being composted and digested in 2011 (Eurostat 2013a, b)

(Scientific-Research Economic Institute of the Ministry of Economy of Belarus 2012).

Animal and plant waste is widely used in farming and woodworking industry (Ministry of Natural Resources and Environmental Protection of the Republic of Belarus 2013b).

6.4.2.2 Estonia

In 2007, the share of biodegradable waste in the total amount of waste generated by an average person per year accounted to 36 % (Estonian Environment Information Centre 2010). The municipal waste collection system covers almost 95 % of population. The country does not have incineration facilities, but one is under development (BiPRO 2012a). The problem of disposal of biodegradable municipal waste is highlighted in the National Waste Management Plan (2008–2013) and the Estonian Environmental Strategy 2030 (Bremere 2011). Experts note that it is very hard quickly and efficiently to introduce separate collection of kitchen waste because the population is not used to it. It would take the investment of time and money in public awareness campaigns to develop and reach optimal levels of kitchen waste (food residues and food waste) collection in households (Zhechkov and Viisimaa 2008).

6.4.2.3 Germany

In 2010, the share of kitchen waste (food residues and food waste) from bio-waste bins amounted to 48 %, the rest comprised composed green waste (Kern et al. 2012). The system of separate collection and treatment of bio-waste (kitchen with a proportion of garden waste) and green waste is one of the most advanced in Europe. Financial incentives for separate collection are included in the waste and charges statutes. However, the green and bio- (kitchen waste with a proportion of garden waste) wastes are collected together (Kern et al. 2012). In addition, the combined collection systems of these two types of waste have not been introduced universally throughout the country. There are 96 municipalities (out of 405) which do not provide their residents with any bio-waste bin. However, starting from the year 2015, a separate collection of bio-waste (kitchen waste with a proportion of garden waste) will become mandatory (Kern et al. 2012).

The government widely supports usage of bio-waste (kitchen waste with a proportion of garden waste) for production of energy. According to the new German Renewable Energy Sources Act (EEG 2012), electricity, generated by facilities using biogas, produced by anaerobic digestion of bio-waste, attracts a higher subsidy rate than if the biogas is produced by digesting other types of biomass (Kern et al. 2012).

Today, nearly 1000 composting plants with a total capacity of more than 10 million metric tonnes are operating in Germany. Half of these plant process exclusively green waste and half treating both bio-waste (kitchen with a proportion of garden waste) and green waste (Kern et al. 2012). Moreover, there are several thousand anaerobic digestion plants, most of which are intended as agricultural facilities for fermenting liquid manure and energy crops. Digestion capacity for bio-waste (kitchen with a proportion of garden waste) is still being built up. At the beginning of the year 2012 around 100 sheer bio-waste digestion plants were in operation (Kern et al. 2012).

6.4.2.4 Latvia

The Latvian Waste Management Law prohibits the disposal of waste from the food industry. The country has started pilot projects on bio-waste treatment (collection and composting), separate collection infrastructure and sustainable waste management are under development (Aleksic 2013). Separate waste collection is promoted by granting exemption from the natural resources tax regarding environmentally harmful goods and packaging (Ministry of Environmental Protection and Regional Development of the Republic of Latvia 2009). Companies that got such exemption are obliged to provide information and educational activities at least four times per year to the general public regarding environmentally sound management of waste (BiPRO 2012b).

6.4.2.5 Lithuania

Presently, the municipal waste collection system covers 94 % of population (ECN 2013). In 2010, 81 % of the amount of biodegradable waste generated in 2000 was landfilled (Kallay 2013). The country has modernised waste collection infrastructure (e.g. trucks, collection bins for separate collection of municipal waste) and initiated public education on waste management (ECN 2013). Furthermore, Lithuania has built 13 green waste composting facilities and another 40 were planned to be built with usage of the EU structural support funds for the period 2007–2013 (EEA 2010). The country set two types of charges for the municipal waste management: fee and local tax and implemented the PAYT schemes at the municipal level (ECN 2013). Currently, 21 green waste composting sites and 157 899 composing containers (boxes) for home composting of biodegradable waste are used in the country (ECN 2013).

6.4.2.6 Poland

Bio-waste is a strongly dominated fraction in household waste from Polish cities, followed by paper/cardboard and plastics (Boer et al. 2010). In the total mass of municipal waste generated in Poland in 2008, a fraction of bio-waste constituted 54.7 % (Deloitte Poland, Fortum, 4P research mix 2011). In 2008, the largest

Table 6.2 Share of kitchen and garden waste in municipal waste generated in Poland in 2008 (Polish Council of Ministers 2010)

	Total (%)	in large cities (>50 k) 37.18 % of polish residents (%)	in small towns (<50 k) 23.89 % of polish residents (%)	in rural areas 38.93 % of polish residents (%)
Kitchen and garden waste, share in total in each category		28.9	36.7	33.1
Kitchen and garden waste, share in total amount of waste	32.1	13.1	9.56	9.5

fraction of kitchen and garden waste in total municipal waste was generated in large cities. However, in small towns Polish residents generated the largest fraction of this type of waste of the total amount of waste generated in this area of residence (Table 6.2).

Experts also predict a continuous increase in the generation of biodegradable municipal waste during period 2011–2022 (Fig. 6.24), as well as of bio-waste from agriculture, horticulture, hydroponic cultivation, forestry, hunting and fishing, food preparation and processing (Fig. 6.25).

6.4.2.7 Sweden

In Sweden, 155 of 290 municipalities are doing bio-waste separate collection, and 62 others are setting up this type of separate collection, in order to use bio-waste anaerobic digestion to produce biogas (Viel 2011). In 2011, 14.9 % of MSW was

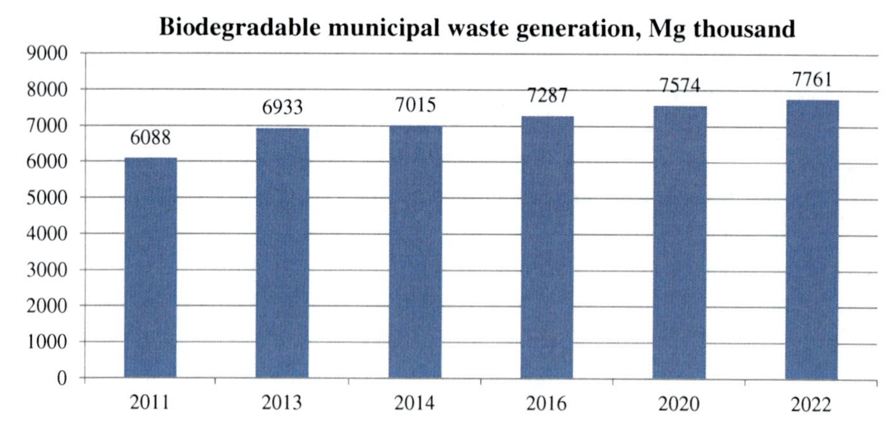

Fig. 6.24 Forecasted amounts of municipal biodegradable waste generated in 2011–2022 (Polish Council of Ministers 2010)

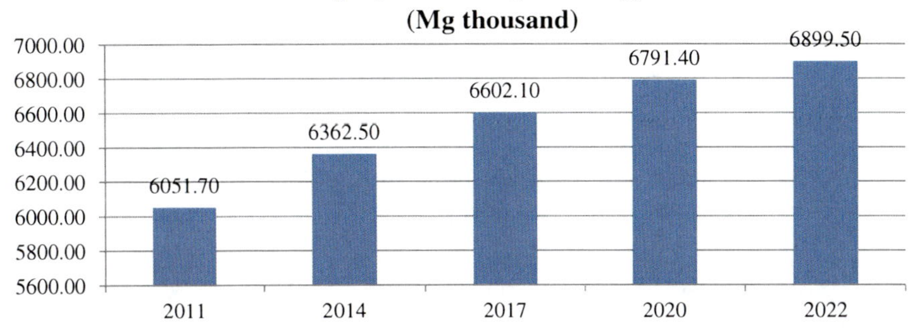

Fig. 6.25 Forecast for production of amount of bio-waste from agriculture, horticulture, hydroponic cultivation, forestry, hunting and fishing, food preparation and processing (Polish Council of Ministers 2010)

biologically treated. It is also important to note that existing capacities for waste incineration are greater than domestic demand for combustible waste (Avfall Sverige 2012).

Thus, the varied states of economic development, renewable energy production, food consumption pattern and biodegradable waste management in the discussed countries, in some way, represent a possible direction of future development of these aspects and, particularly, of the food waste related activities. These give a solid base for the empirical component of this study.

References

Aleksic, D. (2013). *Municipal waste management in Latvia*. Retrieved from http://www.eea.europa.eu/publications/managing-municipal-solid-waste/latvia-municipal-waste-management.

Avfall, S. (2012). *Swedish waste management 2012*. Retrieved from http://www.avfallsverige.se/fileadmin/uploads/Rapporter/SWM2012.pdf.

BalticExport (2013). Economy of Latvia. Retrieved December 22, 2013 from http://balticexport.com/?article=latvijas-ekonomika.

Bank of Lithuania (2013). *Lithuanian economic review*. Retrieved from http://www.lb.lt/lithuanian_economic_review_may_2013#page=5.

BiPRO (2012a). *Country factsheet Estonia (EE)*. Retrieved from http://ec.europa.eu/environment/waste/framework/pdf/EEfactsheet_FINAL.pdf.

BiPRO (2012b). *Country factsheet Latvia (LV)*. Retrieved from http://ec.europa.eu/environment/waste/framework/pdf/LVfactsheet_FINAL.pdf.

Boer, E. D. et al. (2010). A review of municipal solid waste composition and quantities in Poland. *Waste management (New York, N.Y.)*, 30(3), pp.369–77. Retrieved November 26, 2013 from http://www.ncbi.nlm.nih.gov/pubmed/19833497.

Bremere, I. (2011). *Improving waste prevention policy in the Baltic states*. Retrieved from http://www.bef-de.org/Members/befadmin/publikationen/WP2011/activity4-1-1_recommendations_waste-prev.pdf.

Central Statistical Office of Poland (2013). *Foreign trade. I-IX 2013*. Warsaw. Retrieved from http://www.stat.gov.pl/cps/rde/xbcr/gus/pt_foreign_trade_01-09_2013.pdf.

Deloitte Poland; Fortum; 4P research mix (2011). *Waste management in poland challenges in view of EU requirements and legislative changes; public opinion and prospects*. Retrieved from http://www.deloitte.com/assets/Dcom-Poland/LocalAssets/Documents/Raporty,badania,rankingi/pl_Report_WastemanagementinPoland.pdf.

Directive 2009/28/EC (2009). *Directive 2009/28/EC of the European parliament and of the council of 23 April 2009 on the promotion of the use of energy from renewable sources and amending and subsequently repealing Directives 2001/77/EC and 2003/30/EC*. Official Journal L 140, 5/6/2009 pp. 16-62. Retrieved from http://eur-lex.europa.eu/LexUriServ/LexUriServ.do?uri=Oj:L:2009:140:0016:0062:en:PDF.

ECN (2013). Country report of Lithuania. *Organic resources and biological treatment*. Retrieved November 28, 2013 from http://www.compostnetwork.info/country-report-of-lithuania.html.

EEA (2010). Waste (Lithuania). *Lithuania*. Retrieved November 7, 2013 from http://www.eea.europa.eu/soer/countries/lt/soertopic_view?topic=waste.

ERS/USDA (2013). Food expenditures. Retrieved September 12, 2013 from http://www.ers.usda.gov/data-products/food-expenditures.aspx#.UrrPO_QW1dN.

Esko, T. et al. (2012). *FOODWEB consumer awareness study report II*. Retrieved from http://foodweb.ut.ee/s2/111_94_86_FOODWEB_Consumers_Awareness_Study_report_II_Eston.pdf.

Estonian Environment Information Centre (2010). *Estonian environmental review 2009*. Tallinn. Retrieved from http://www.keskkonnainfo.ee/publications/4263_PDF.pdf.

Estonia.eu (2013). Economy in numbers. Retrieved November 14, 2013 from http://estonia.eu/about-estonia/economy-a-it/economy-in-numbers.html.

Estonian Ministry of Agriculture (2010). Agriculture and food market. Retrieved November 17, 2013 from http://www.agri.ee/agriculture-and-food/.

European Commission (2013a). Agriculture in the European Union and the Member States—Statistical factsheets. *Statistics and indicators*. Retrieved December 7, 2013 from http://ec.europa.eu/agriculture/statistics/factsheets/.

European Union (2013). Member countries of the European Union. Retrieved November 12, 2013 from http://europa.eu/about-eu/countries/member-countries/index_en.htm.

Eurostat (2013a). Children at risk of poverty or social exclusion. Retrieved December 22, 2013 from http://epp.eurostat.ec.europa.eu/statistics_explained/index.php/Children_at_risk_of_poverty_or_social_exclusion.

Eurostat (2013b). *Environment in the EU27*, (March) (pp. 1–3). Retrieved from http://epp.eurostat.ec.europa.eu/cache/ITY_PUBLIC/8-04032013-BP/EN/8-04032013-BP-EN.PDF.

FAO (2013a). *FAO statistical yearbook 2013. World food and agriculture*. Rome. Retrieved from http://www.fao.org/docrep/018/i3107e/i3107e00.htm.

Fischer, C. (2013a). *Municipal waste management in Estonia*. Retrieved from http://www.eea.europa.eu/publications/managing-municipal-solid-waste/estonia-municipal-waste-management.

Fischer, C. (2013b). *Municipal waste management in Germany*. Retrieved from http://www.eea.europa.eu/publications/managing-municipal-solid-waste/germany-municipal-waste-management.

German Federal Statistical Office (2013a). *Environment. Waste balance 2011*. Wiesbaden. Retrieved from https://www.destatis.de/EN/FactsFigures/NationalEconomyEnvironment/Environment/EnvironmentalSurveys/WasteManagement/Tables/WasteBalance2011.pdf?__blob=publicationFile.

German Federal Statistical Office (2013b). Foreign trade. Retrieved December 22, 2013 from https://www.destatis.de/EN/FactsFigures/NationalEconomyEnvironment/ForeignTrade/TradingGoods/Tables/ImportsExports.html.

Gosiewska, M. (2013). *FoRWaRD regional report*. Retrieved from http://foodrecoveryproject.eu/wp-content/uploads/2012/11/Regional-report-Poland.pdf.

Guziana, B. et al. (2012). *Waste-to-Energy in a polish perspective*. Retrieved from http://www.diva-portal.org/smash/get/diva2:574913/FULLTEXT01.pdf.
Index Mundi (2013). Belarus Economy—overview. *Economy*. Retrieved November 12, 2013 from http://www.indexmundi.com/belarus/economy_overview.html.
Institute for Private Finances Swedbank (2013). *Food basket study in Baltic countries*, Available at: http://www.slideshare.net/SwedbankAB/food-basket-study-in-baltic-countries.
Kallay, T. (2013). *Municipal waste management in Lithuania*. Retrieved from http://www.eea.europa.eu/publications/managing-municipal-solid-waste/lithuania-municipal-waste-management.
Kern, M. et al. (2012). *Ecologically sustainable recovery of Bio-Waste*. Berlin. Retrieved from http://www.bmu.de/fileadmin/Daten_BMU/Pools/Broschueren/bioabfaelle_2012_en_bf.pdf.
Lauri, M. (2012). Structure of the economy. *Estonica*. Retrieved from http://www.estonica.org/en/Economy/General_overview_of_Estonian_economy/Structure_of_the_economy/.
London Chamber of Commerce and Industry (2010). *Lithuania*. Retrieved from http://www.londonchamber.co.uk/docimages/7016.pdf.
McKinsey & Company (2012). *Growth and renewal in the Swedish economy*. Retrieved from http://www.mckinsey.com/insights/europe/growth_and_renewal_in_the_swedish_economy.
Milios, L. (2013). *Municipal waste management in Sweden*. Retrieved from http://www.eea.europa.eu/publications/managing-municipal-solid-waste/sweden-municipal-waste-management.
Ministry of Economics of the Republic of Latvia (2013). Trade statistics. Retrieved December 22, 2013 from http://www.em.gov.lv/em/2nd/?lng=en&cat=30288.
Ministry of Economy of the Republic of Poland (2012). *Poland 2012—Economy report*. Warsaw. Retrieved from http://www.mg.gov.pl/files/upload/9142/RoG_20121019_MG_eng_popr.pdf.
Ministry of Environmental Protection and Regional Development of the Republic of Latvia (2009). *Environmental policy strategy 2009–2015*. Riga, Latvia. Retrieved from http://www.varam.gov.lv/eng/dokumenti/politikas_planosanas_dokumenti/.
Ministry of Natural Resources and Environmental Protection of the Republic of Belarus (2010). *The state of environment in the republic of Belarus*. Minsk. Retrieved from http://www.nsmos.by/tmp/fckimages/GIATSzakon/Nats_doklad_eng.pdf.
Ministry of Natural Resources and Environmental Protection of the Republic of Belarus (2013b). Waste Treatment. Retrieved December 1, 2013 from http://minpriroda.gov.by/en/areas/waste.
National Statistical Committee of the Republic of Belarus (2013a). Energy Statistics (Jenergeticheskaja statistika). Retrieved December 22, 2013 from http://belstat.gov.by/homep/ru/indicators/energy.php.
National Statistical Committee of the Republic of Belarus (2013b). Sample Household Living Standards Survey. Retrieved September 12, 2013 from http://belstat.gov.by/homep/en/households/main2.php.
Official Website of the Republic of Belarus (2013). Key facts about Belarus. Retrieved November 14, 2013 from http://www.belarus.by/en/about-belarus/key-facts.
Polish Council of Ministers (2010). National waste management plan 2014., pp.1–83. Retrieved from http://www.mos.gov.pl/artykul/3340_krajowy_plan_gospodarki_odpadami_2014/21693_national_waste_management_plan_2014.html.
Research Center of the Institute for Privatization and Management (2013). *Poverty and vulnerable groups in Belarus—2013 (Bednost' i social'no ujazvimye gruppy v Belarusi — 2013)*. Minsk, Belarus. Retrieved from http://www.research.by/webroot/delivery/files/poverty2013r.pdf.
Rudzitis, E. et al. (2013). *Baltic household outlook*. Retrieved from http://www.seb.lv/data/Analitiska-Info/Makroekonomika/en/Baltic-Household-Outlook_2013-10-ENG.pdf.
Scientific-Research Economic Institute of the Ministry of Economy of Belarus (2012). *Sustainable development of the republic of Belarus based on "green" economy principles*. Minsk. Retrieved from http://undp.by/f/file/green-economy-belarus-en.pdf.
Statistics Estonia 2013. Every fifth person in Estonia lived in relative poverty and every fourteenth in absolute poverty last year. Retrieved December 22, 2013 from http://www.stat.ee/65388.
Swedbank Macro Research (2013). *Swedbank analysis. What drives inflation in the Baltic countries?*. Stockholm. Retrieved from http://www.swedbank.lt/lt/previews/get/3804/rss.

Tylaite, K. & Bastys, M. (2013). *FoRWaRD regional report*. Retrieved from http://foodrecoveryproject.eu/wp-content/uploads/2012/11/Regional_report-Lithuania.pdf.
Viel, D. (2011). *Food wastage study mid-term report*. Retrieved from http://www.developpement-durable.gouv.fr/IMG/food waste mid-term report_VF.pdf.
Vorne, V. et al. (2011). *The Baltic environment, food and health: from habits to awareness. Feasibility study*. Retrieved from http://foodweb.ut.ee/s2/111_94_42_MTT_Report_34_The_Baltic_environment_food_and_he.pdf.
World Bank (2013a). Depth of the food deficit (kilocalories per person per day). Retrieved November 22, 2013 from http://data.worldbank.org/indicator/SN.ITK.DFCT.
World Bank (2013b). Prevalence of undernourishment (% of population). Retrieved December 22, 2013 from http://data.worldbank.org/indicator/SN.ITK.DEFC.ZS.
Zhechkov, R. & Viisimaa, M. (2008). *Evaluation of waste policies related to the landfill directive*. Copenhagen. Retrieved from http://scp.eionet.europa.eu/publications/wp2008_3/wp/wp2008_3.

Chapter 7
The State of the Problem of Food Waste in the Baltic Region Countries

Due to the limited available information on food waste in Baltic countries, this book undertakes an extensive research based on appraisals of information available at various sources, and compiles the data into various blocks. This chapter presents the state-of-the-art of the problem of food waste across the Baltic Sea Region and methods of its treatment.

7.1 Food Waste Generation in the Baltic

Today, among a limited number of the research works regarding the problem of food waste in the EU countries, the results of the 'Preparatory study on food waste across EU 27' (BIO Intelligence Service 2010), conducted by BIO Intelligence Service for the European Commission, are referred and cited mostly. According to the study, the biggest share of food waste among six discussed countries is generated in Germany and Poland, 11.63 % and 10.05 % respectively (Fig. 7.1), in weight terms it amounts to 10.387 and 8.972 thousand tonnes (Table 7.1). However, in this particular case, the term 'food waste' includes all types of food related waste except agricultural waste. In the current work, mix of these waste types is defined as food wastage (Fig. 1.1).

According to the study, a 'leading' sector in food wastage generation varies with each country. Mainly, the results could be divided into two groups. The first group are countries, such as Germany and Sweden, where the largest amount is generated by households. The second group are countries, except Lithuania, where most of waste is generated by the manufacturing sector (Fig. 7.2).

With regard to the amount of food wastage generated by the manufacturing sector, the study outlines that most of it is unavoidable waste such as food losses, by-products and food residues. The share of food waste in household waste is 25 %, the value was taken from the studies conducted by WRAP (BIO Intelligence Service 2010). However, the results of the WRAP studies represent the situation in the UK and might also be applied to other countries with a similar economic and social situation, consumption pattern and behaviour as Germany and Sweden. But even if

W. Leal Filho and M. Kovaleva, *Food Waste and Sustainable Food Waste Management in the Baltic Sea Region*, Environmental Science and Engineering, DOI 10.1007/978-3-319-10906-0_7

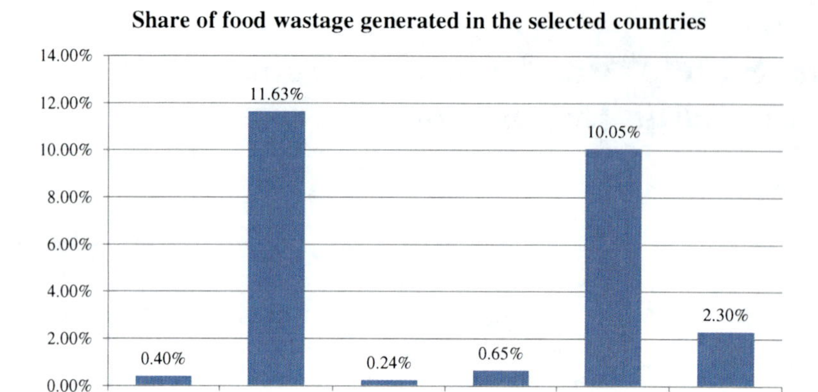

Fig. 7.1 Percentage of food wastage generated in the selected countries in the total amount of food wastage generated by the EU Member States (BIO Intelligence Service 2010)

Table 7.1 Total food wastage generation in the selected countries: Best estimate by Member State, in tonnes (BIO Intelligence Service 2010)

Countries	Manufacturing	Households	Other sectors[a]	Total
Estonia	237,257	82,236	36,000	355,000
Germany	1,848,881	7,676,471	862,000	10,387,000
Latvia	125,635	78,983	11,000	216,000
Lithuania	222,205	111,160	248,000	581,000
Poland	6,566,060	2,049,844	356,000	8,972,000
Sweden	601,327	905,000	547,000	2,053,000

[a] Other sectors includes Wholesale/Retail and Food Service/Catering sectors

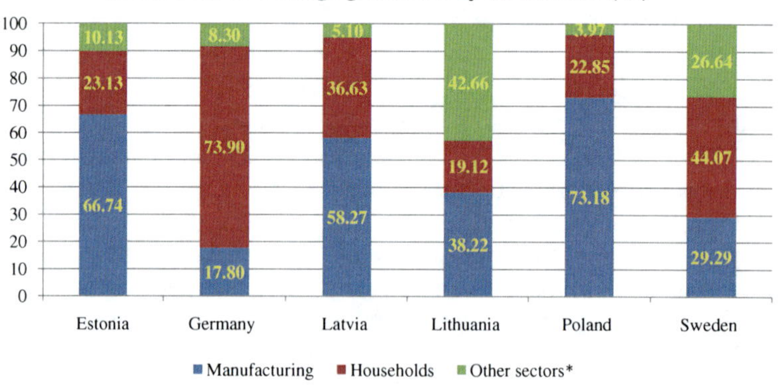

Fig. 7.2 Percentage share of food wastage generated by each sector out of total food wastage generation in the selected countries (BIO Intelligence Service 2010) * Under Other sectors authors undermined Wholesale/Retail sector and Food Service/Catering sector

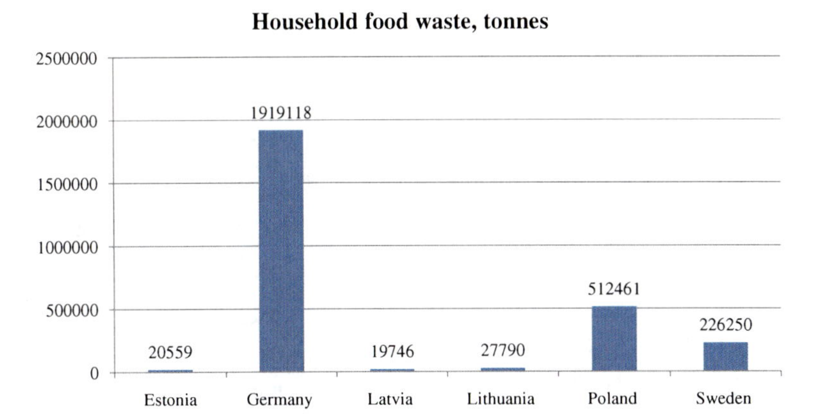

Fig. 7.3 The amount of food waste generated by households (25 % of the total amount of household food wastage) (BIO Intelligence Service 2010)

we assume that the share of household food waste in all countries is equal to 25 %, the calculations show that Germany generates the biggest amount (1,919,118 tonnes) and it is almost four times bigger than the amount of food waste generated by Poland, who follow (512,461 tonnes) (Fig. 7.3).

It is interesting to compare these results with the results of the survey conducted by Eurobarometer (2011). The table below shows the percentage of food that respondents in different countries think they throw away (Table 7.2).

An analysis of the results of both studies shows a curious relationship between actual waste amounts and people's perception of the amount of waste that they generate. Respondents in the countries where the share of household food wastage is the biggest think that they throw least amounts of food. For example, in Germany and Sweden, 81 % and 77 % of the respondents think that they throw <15 % of food (Table 7.2).

Apart from Eurostat that presents data on different types of food related waste, another source for such types of waste is the FAO food balance sheets.

Table 7.2 Estimated percentage of food that goes to waste in the selected countries (Eurobarometer 2011)

Countries	More than 30 %	16–30 %	15 % or less	None	NA
Germany	3	11	81	5	0
Sweden	2	14	77	5	2
Latvia	6	17	64	12	1
Lithuania	7	17	55	19	2
Poland	1	12	67	21	0
Estonia	4	10	62	21	1

7.1.1 Food Waste Amounts According to the FAO Food Balance Sheets

A country's food balance sheet includes data on wastage amounts that defined as lost at all stages between the level at which production is recorded and the household, i.e. losses during storage and transportation, marketing stages (the lack of adequate marketing systems and organization, the imbalances of supply and demand), as well as technical losses occurring during the transformation of the primary commodities into processed products. It excludes losses occurring during the pre-harvest and harvesting stages as well as both edible and inedible parts of the commodity occurring in the household, e.g. in the kitchen (FAO Economic and Social Development Department 2001). Thus the amounts discussed below include avoidable and unavoidable fractions.

The latest data are available for the year 2009. An analysis of the data brings a number of conclusions. Firstly, the results of the comparison of the amounts of wastage generated by the discussed countries in the absolute terms differ from the results of the comparison of the relative terms. In other words, for example, in absolute terms Germany and Poland generated the highest amounts of fruit wastage among six countries, 346 and 222 thousand tonnes respectively (Fig. 7.4). Whereas, the comparison of the amounts of fruit wastage generated per one hundred tonnes of total available supply shows that the highest amounts were generated by Poland and Belarus (Fig. 7.5). The same pattern can be found in an analysis of the amounts of generated banana wastage (Figs. 7.6 and 7.7).

In addition to banana wastage, the amount of fruit wastage includes among others grape, lemon, lime, apple, orange, mandarine and pineapple wastages. The biggest generators of these types of wastage are Germany and Poland (Figs. 7.8, 7.9, 7.10, 7.11 and 7.12).

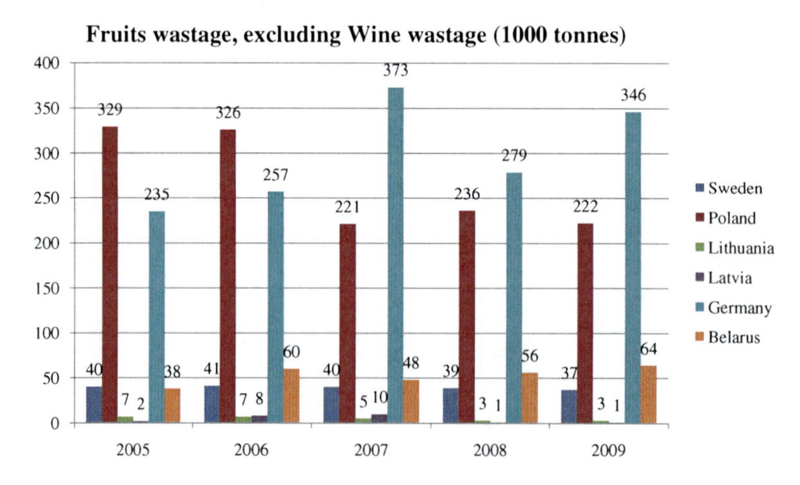

Fig. 7.4 Fruits wastage, excluding Wine wastage (1000 tonnes) generated in the selected countries in 2005–2009 (FAO 2013)

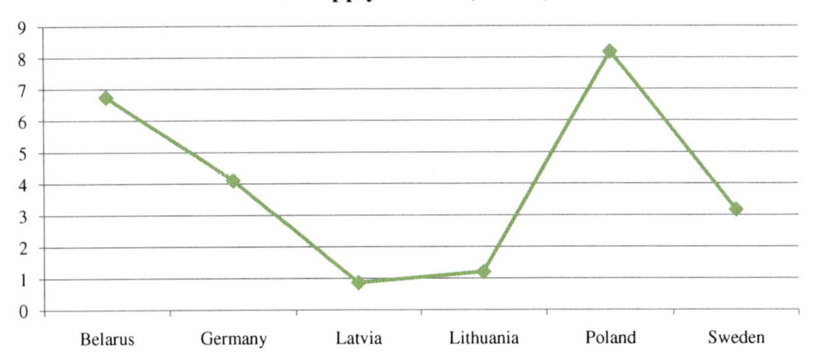

Fig. 7.5 Amount of fruit wastage, excluding wine wastage, generated by the selected countries per one hundred tonnes of fruit total supply, in tonnes in 2009 (FAO 2013a)

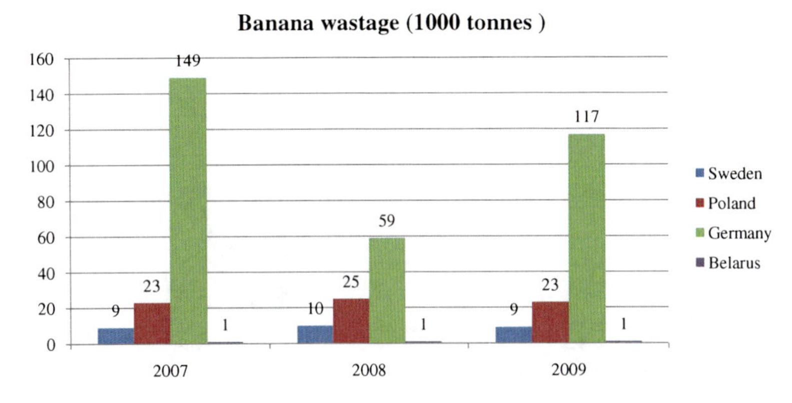

Fig. 7.6 Banana wastage (1000 tonnes) generated in 2007–2009 (FAO 2013a)

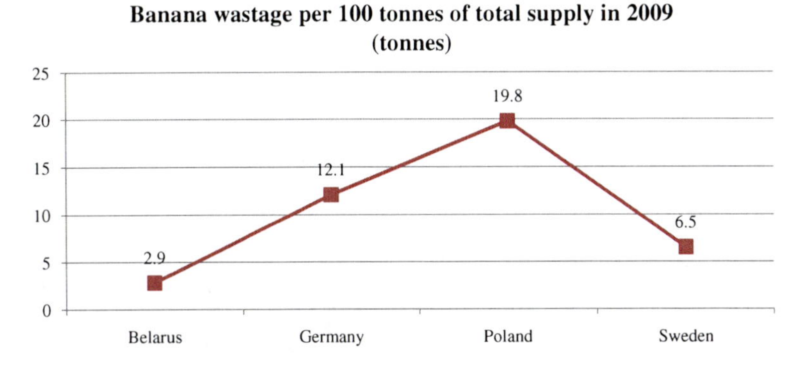

Fig. 7.7 Amount of Banana wastage generated by the selected countries per one hundred tonnes of banana total supply, in tonnes in 2009 (FAO 2013a)

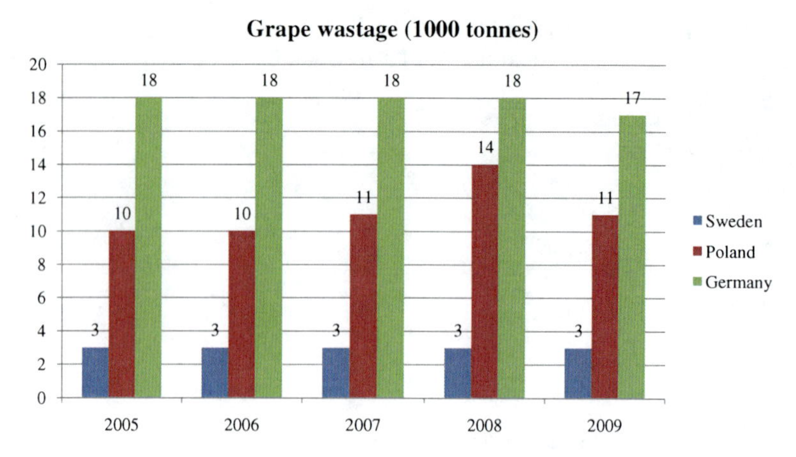

Fig. 7.8 Grape wastage (1000 tonnes) generated in the selected countries in 2005–2009 (FAO 2013a)

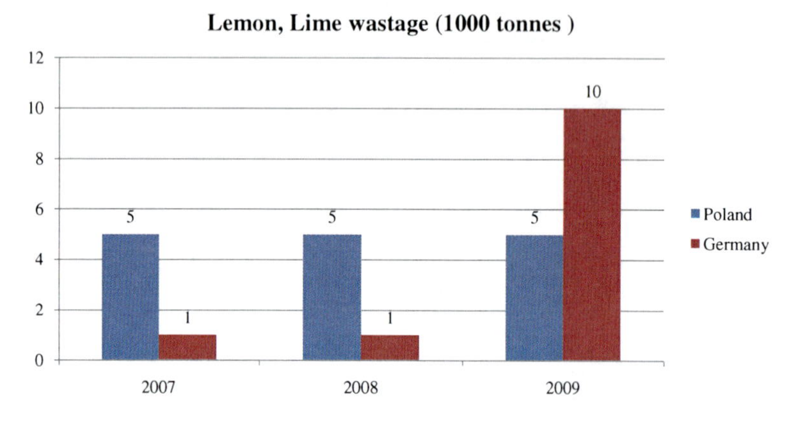

Fig. 7.9 Lemon, Lime wastage (1000 tonnes) generated in Poland and Germany in 2007–2009 (FAO 2013a)

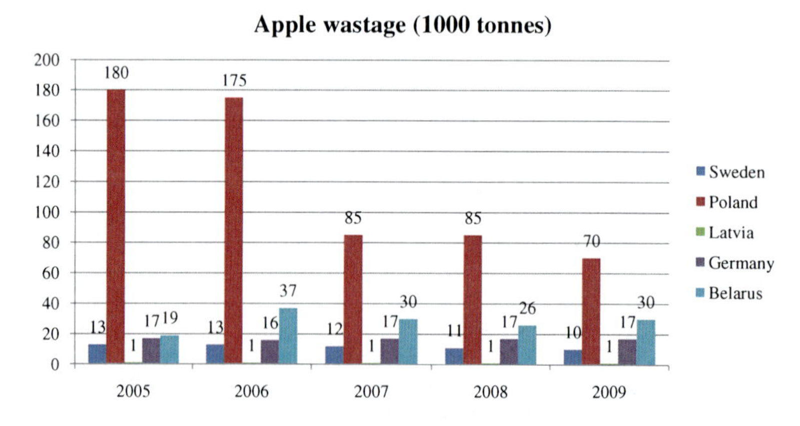

Fig. 7.10 Apple wastage (1000 tonnes) generated by the selected countries in 2005–2009 (FAO 2013a)

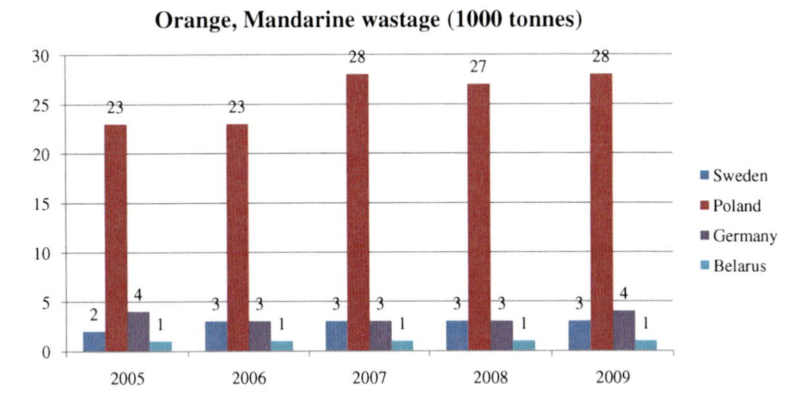

Fig. 7.11 Orange, Mandarine wastage (1000 tonnes) generated in the selected countries in 2005–2009 (FAO 2013a)

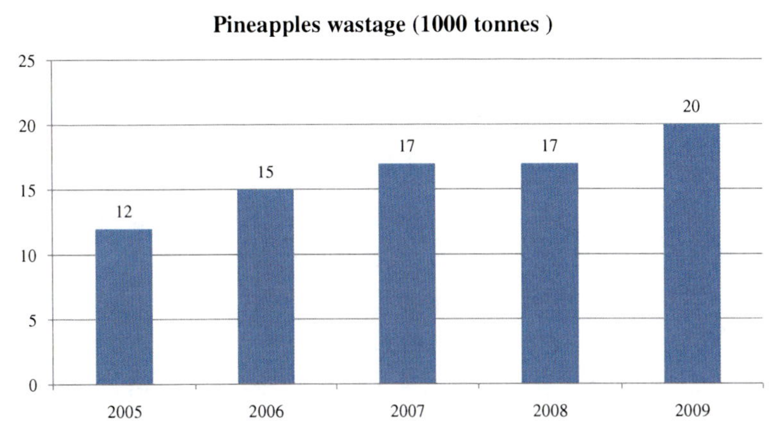

Fig. 7.12 Pineapple wastage (1000 tonnes) generated in Germany in 2005–2009 (FAO 2013a)

The amount of meat wastage includes bovine, poultry, pig, mutton and goat, and other meat wastage types (Figs. 7.15, 7.16 and 7.17). It is important to note that the data for the amount of meat wastage are not available for all seven countries. In absolute terms, between years 2006 and 2008, Estonia and in 2009, Germany and Poland generated the biggest amounts of meat wastage (Fig. 7.13). At the same time, the comparison of the amount generated per one hundred tonnes shows that in 2009, Estonia, followed by Lithuania and Latvia generated the biggest amounts of meat wastage among six countries (Fig. 7.14).

The amount of cereal wastage includes such types as wastage as wheat, rice, barley, maize, rye, oats, millet, sorghum and others (Figs. 7.20, 7.21, 7.22, 7.23, 7.24 and 7.25). According to the FAO, between years 2005 and 2009, in relative and absolute terms, Germany and Poland generated the biggest amount of cereal wastage (Figs. 7.18 and 7.19).

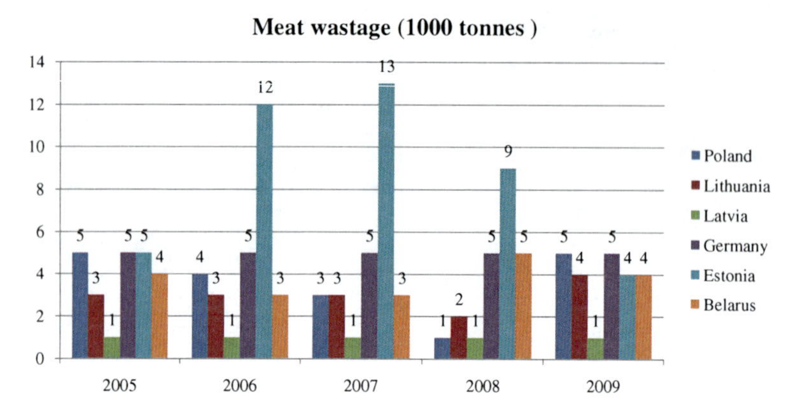

Fig. 7.13 Meat wastage (1000 tonnes) generated in the selected countries in 2005–2009 (FAO 2013a)

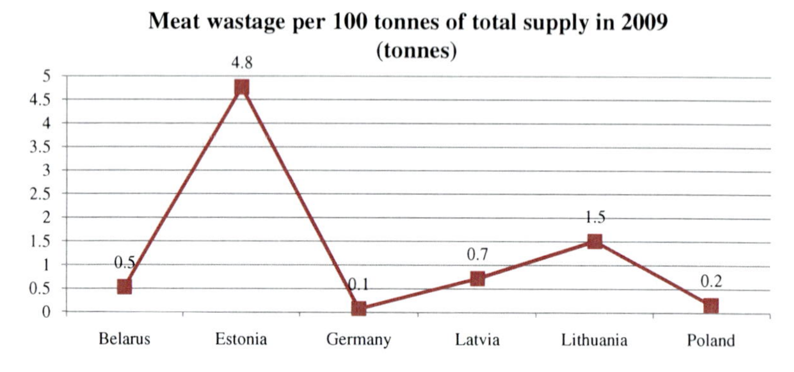

Fig. 7.14 Amount of Meat wastage generated by the selected countries per one hundred tonnes of meat total supply in 2009 (FAO 2013a)

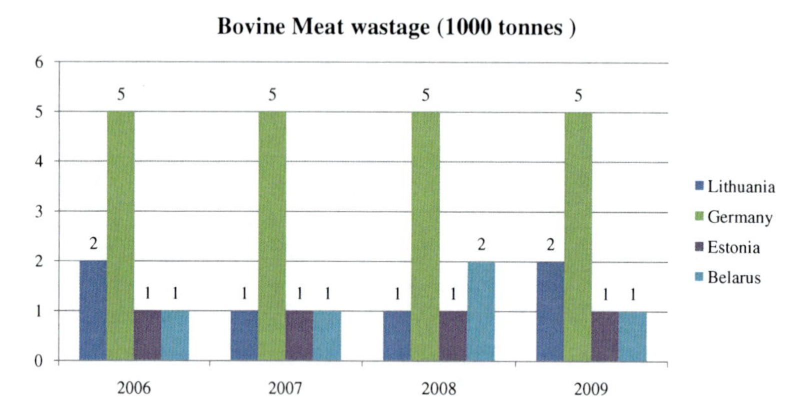

Fig. 7.15 Bovine Meat wastage (1000 tonnes) generated in the selected countries in 2006–2009 (FAO 2013a)

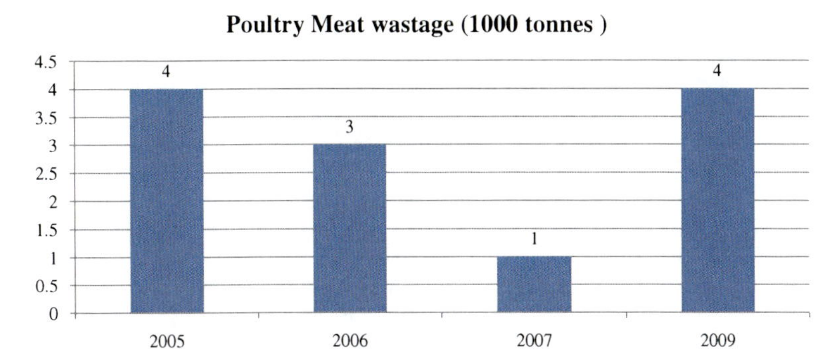

Fig. 7.16 Poultry Meat wastage (1000 tonnes) generated in Poland in 2005–2009 (FAO 2013a)

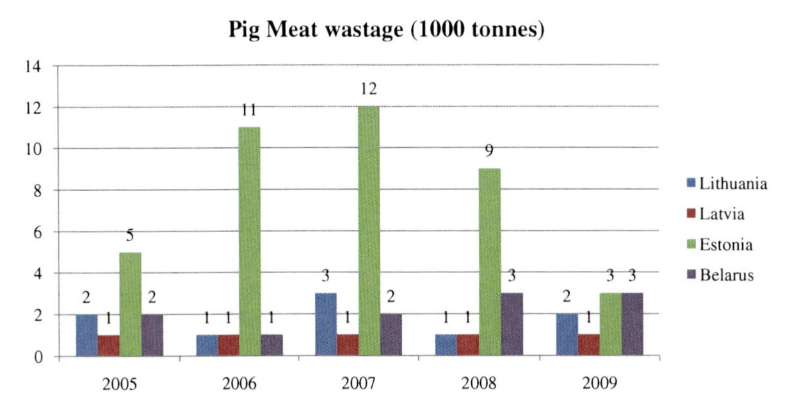

Fig. 7.17 Pig Meat wastage (1000 tonnes) generated in the selected countries in 2005–2009 (FAO 2013a)

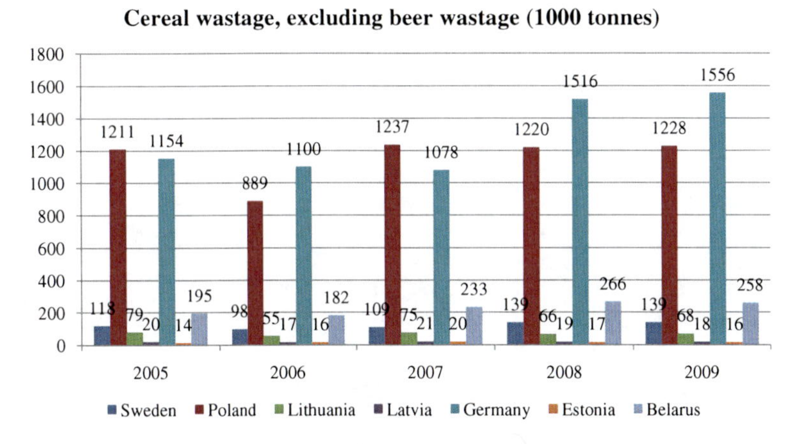

Fig. 7.18 Cereal wastage, excluding beer wastage (1000 tonnes) generated in the selected countries in 2005–2009 (FAO 2013a)

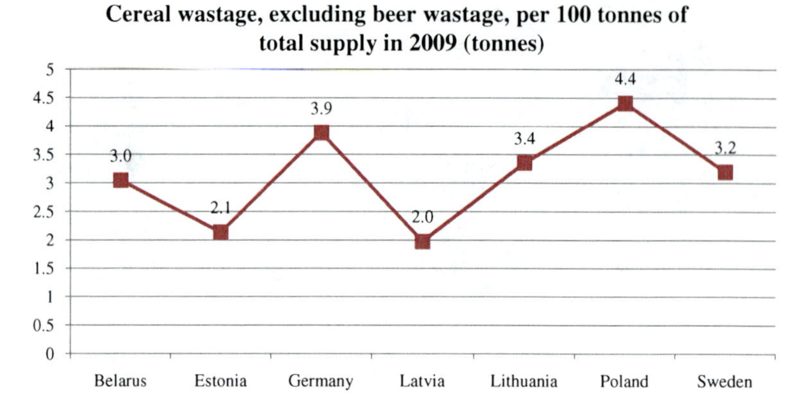

Fig. 7.19 Amount of cereal wastage, excluding beer wastage generated by the selected countries per one hundred tonnes of cereal total supply in 2009 (FAO 2013a)

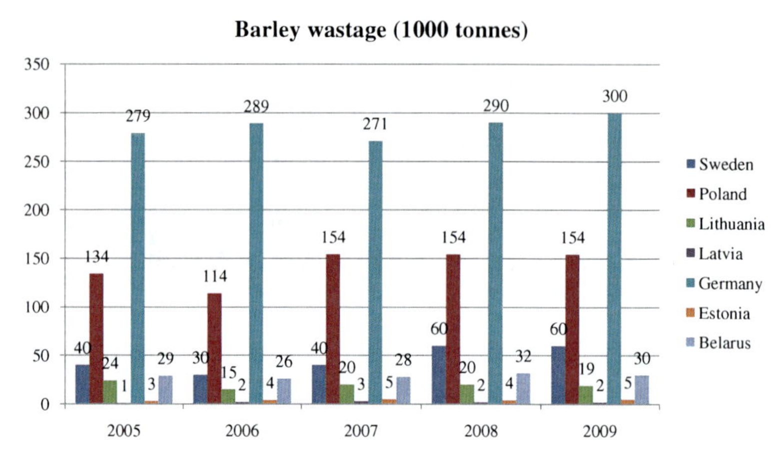

Fig. 7.20 Barley wastage (1000 tonnes) generated in the selected countries in 2005–2009 (FAO 2013a)

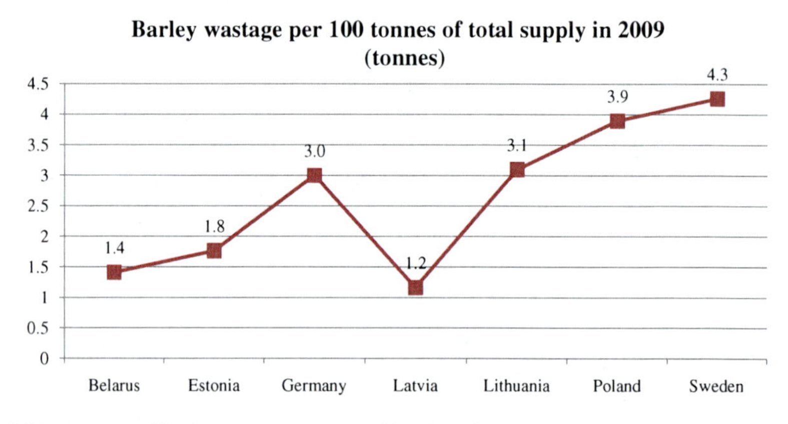

Fig. 7.21 Amount of barley wastage generated by the selected countries per one hundred tonnes of barley total supply in 2009 (FAO 2013a)

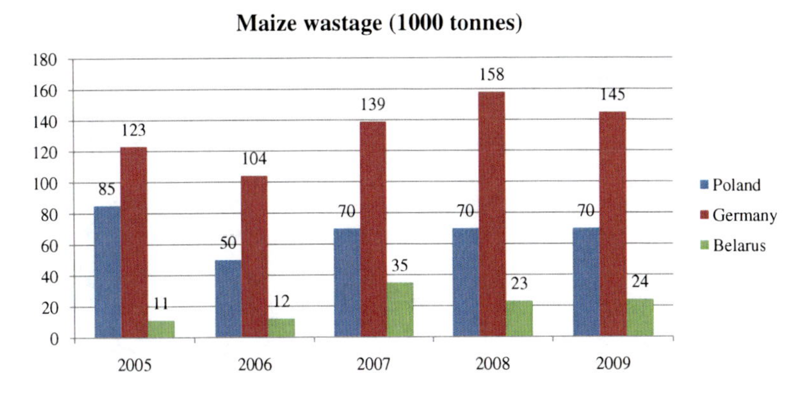

Fig. 7.22 Maize wastage (1000 tonnes) generated in the selected countries in 2005–2009 (FAO 2013a)

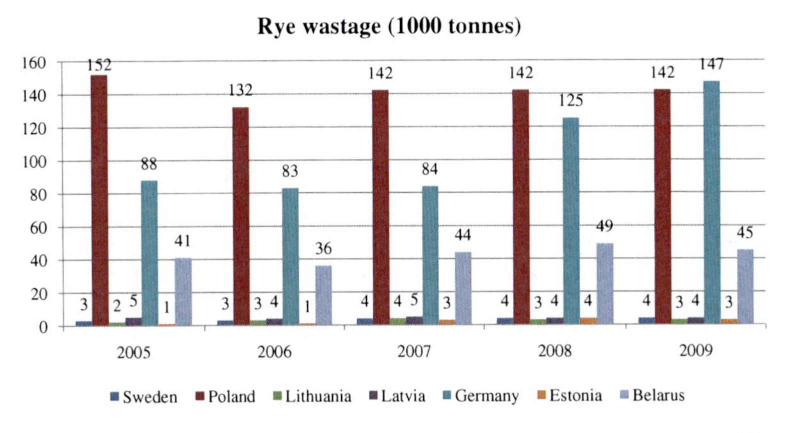

Fig. 7.23 Rye wastage (1000 tonnes) generated in the selected countries in 2005–2009 (FAO 2013a)

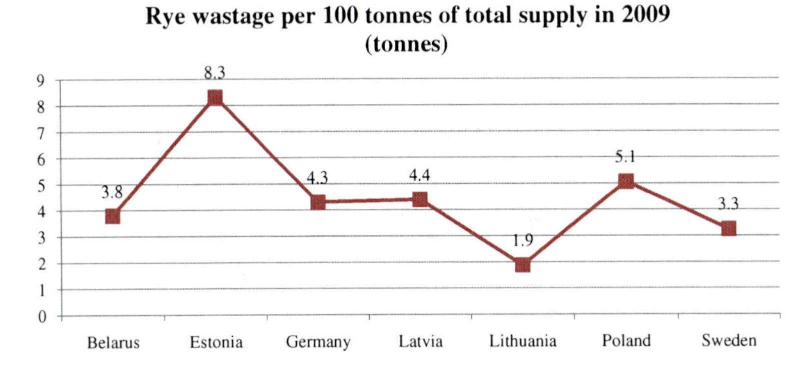

Fig. 7.24 Amount of rye wastage generated by the selected countries per one hundred tonnes of rye total supply in 2009 (FAO 2013a)

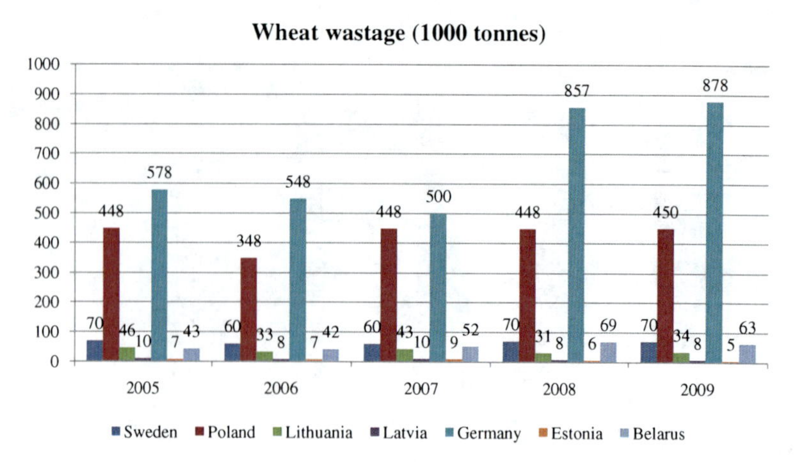

Fig. 7.25 Wheat wastage (1000 tonnes) generated in the selected countries in 2005–2009 (FAO 2013a)

Pulses wastage that includes bean and pea wastage (Figs. 7.28 and 7.29) is mostly generated by Belarus and Poland (Figs. 7.26 and 7.27).

Vegetable wastage includes among other types onion and tomato wastage (Figs. 7.32 and 7.33). The largest amounts are also generated by Germany and Poland (Figs. 7.30, 7.31 and 7.34).

The amount of potato wastage, which refers to the starchy roots group of wastage, according to the FAO classification, mostly generated by Germany and Poland in both absolute and relative terms (Figs. 7.35 and 7.36).

Germany is also a 'leader' in the generation of the highest amount of egg wastage, however, in this example the data are only available for Germany, Poland and Sweden (Fig. 7.37). The largest amount per one hundred tonnes is generated by Sweden (Fig. 7.38).

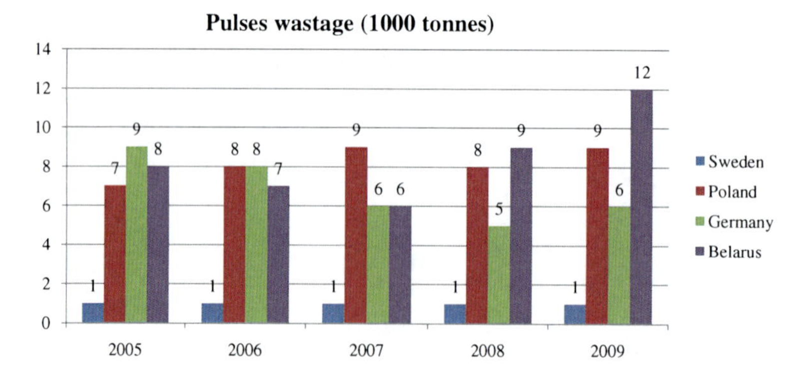

Fig. 7.26 Pulses wastage (1000 tonnes) generated in the selected countries in 2005–2009 (FAO 2013a)

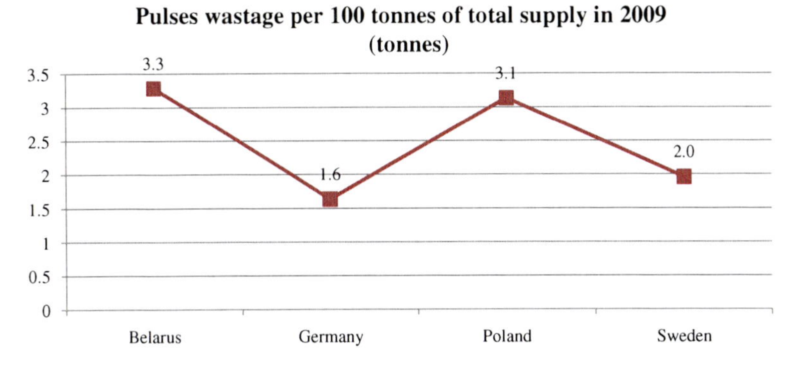

Fig. 7.27 Amount of pulses wasted generated by the selected countries per one hundred tonnes of pulses total supply in 2009 (FAO 2013a)

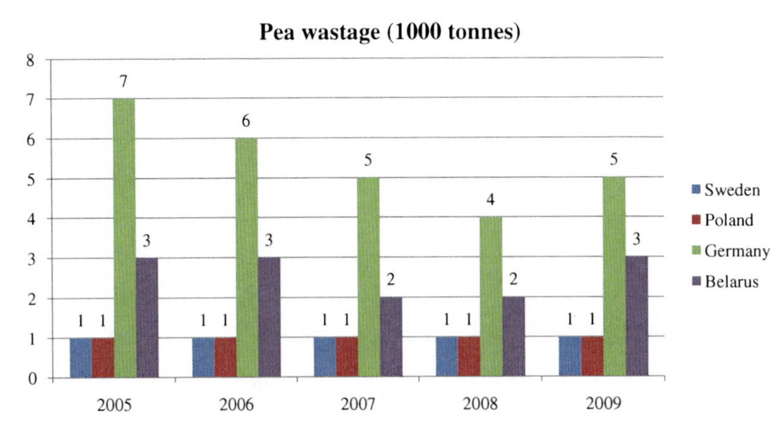

Fig. 7.28 Pea wastage (1000 tonnes) generated in the selected countries in 2005–2009 (FAO 2013a)

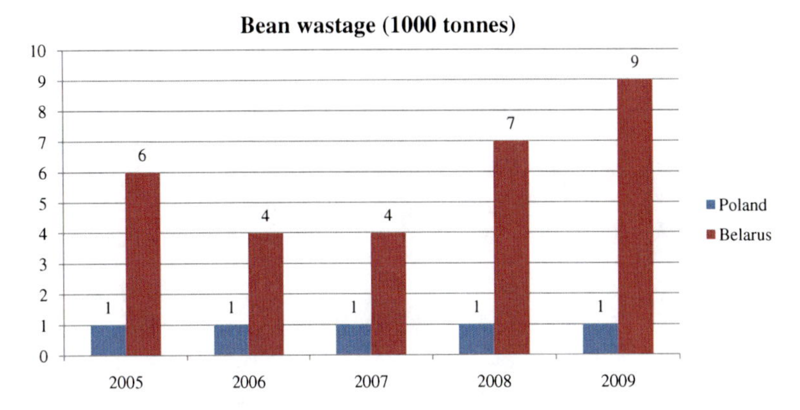

Fig. 7.29 Bean wastage (1000 tonnes) generated by Poland and Belarus in 2005–2009 (FAO 2013a)

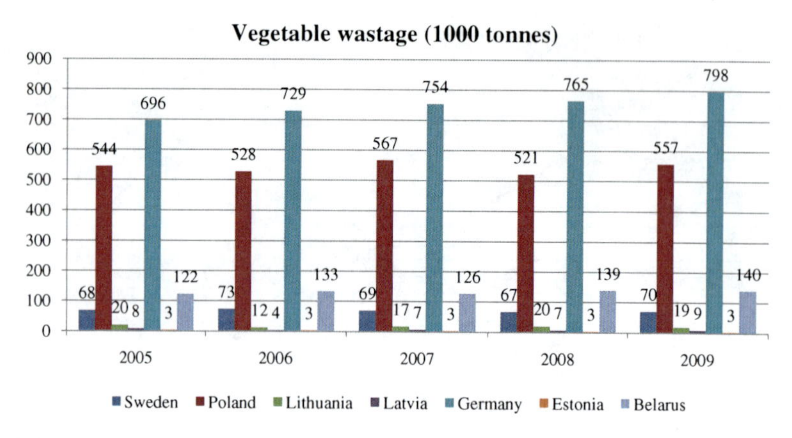

Fig. 7.30 Vegetable wastage (1000 tonnes) generated in the selected countries in 2005–2009 (FAO 2013a)

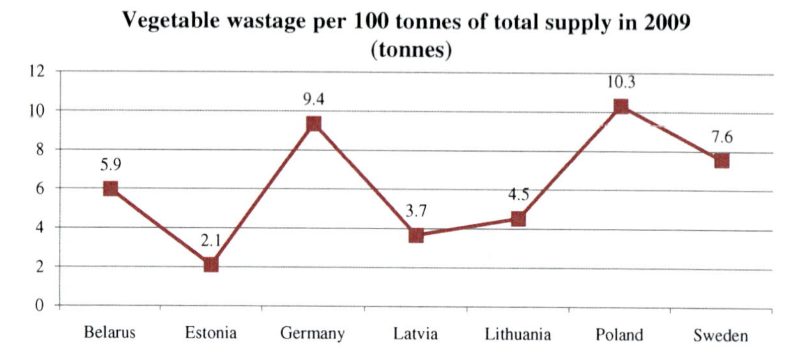

Fig. 7.31 Amount of vegetable wastage generated by the selected countries per one hundred tonnes of vegetable total supply in 2009 (FAO 2013a)

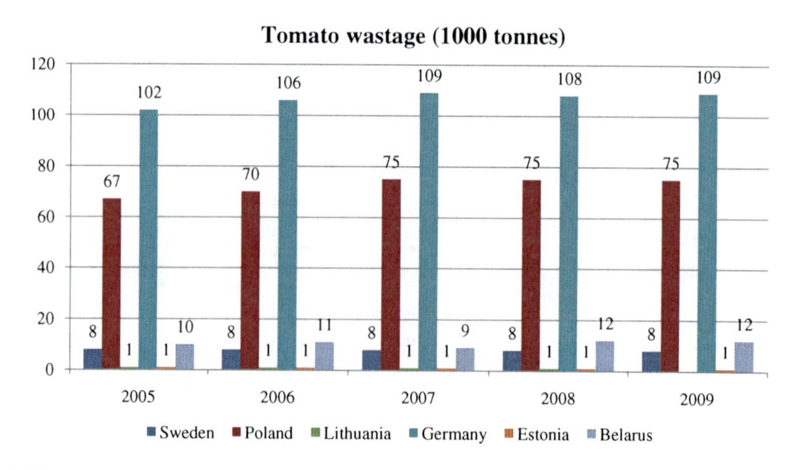

Fig. 7.32 Tomato wastage (1000 tonnes) generated in the selected countries in 2005–2009 (FAO 2013a)

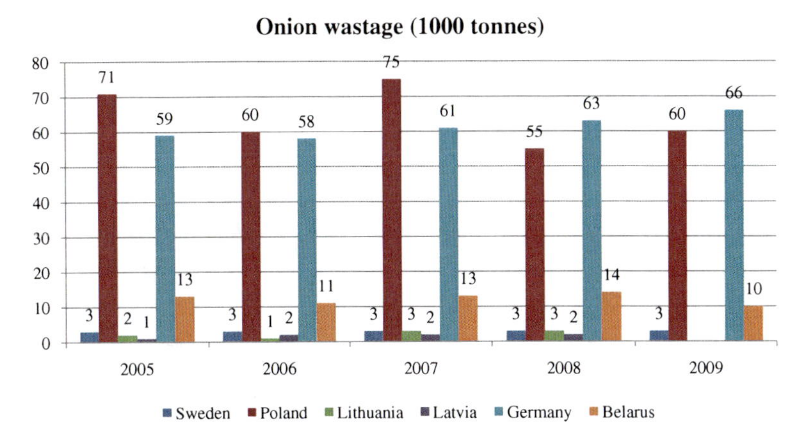

Fig. 7.33 Onion wastage (1000 tonnes) generated in the selected countries in 2005–2009 (FAO 2013a)

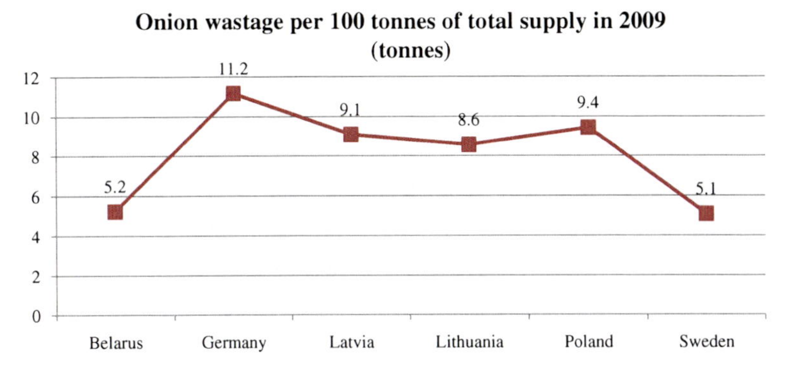

Fig. 7.34 Amount of onion wastage generated by the selected countries per one hundred tonnes of onions total supply in 2009 (FAO 2013a)

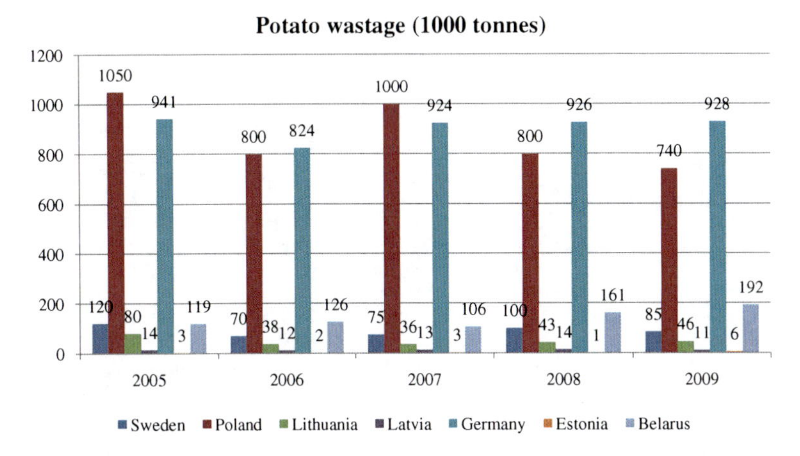

Fig. 7.35 Potato wastage (1000 tonnes) generated in the selected countries in 2005–2009 (FAO 2013a)

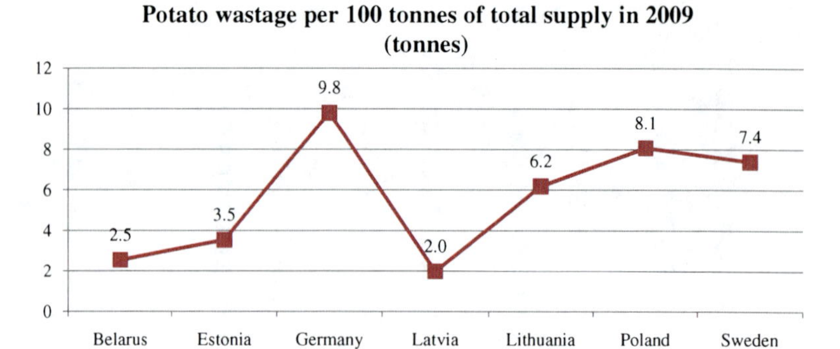

Fig. 7.36 Amount of potato wastage generated by the selected countries per hundred tonnes of potatoes total supply in 2009 (FAO 2013a)

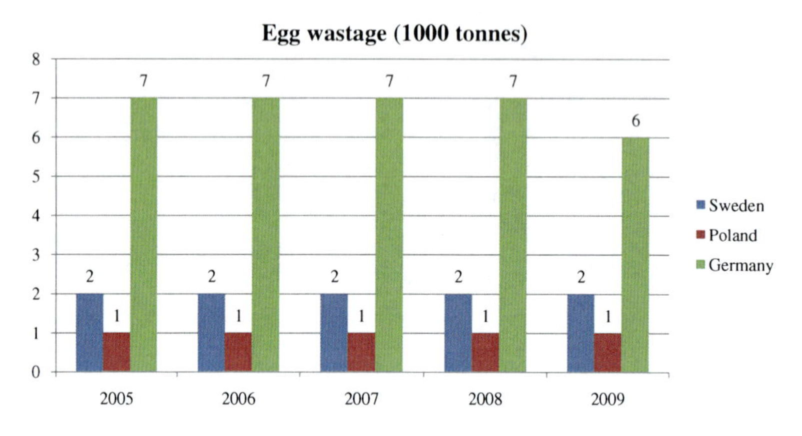

Fig. 7.37 Egg wastage (1000 tonnes) generated in the selected countries in 2005–2009 (FAO 2013a)

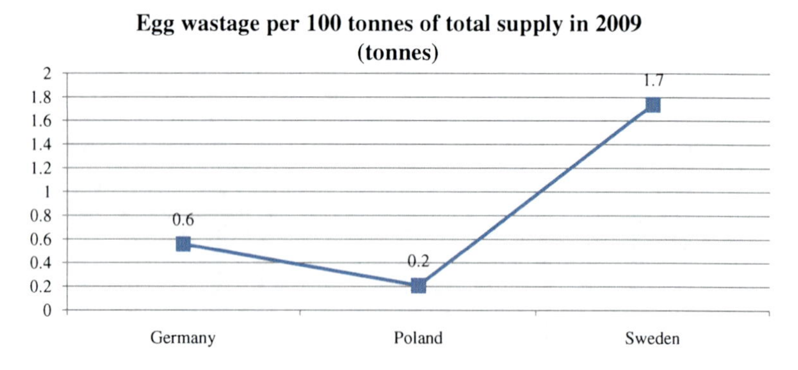

Fig. 7.38 Amount of egg wastage generated by the selected countries per one hundred tonnes of eggs total supply in 2009 (FAO 2013a)

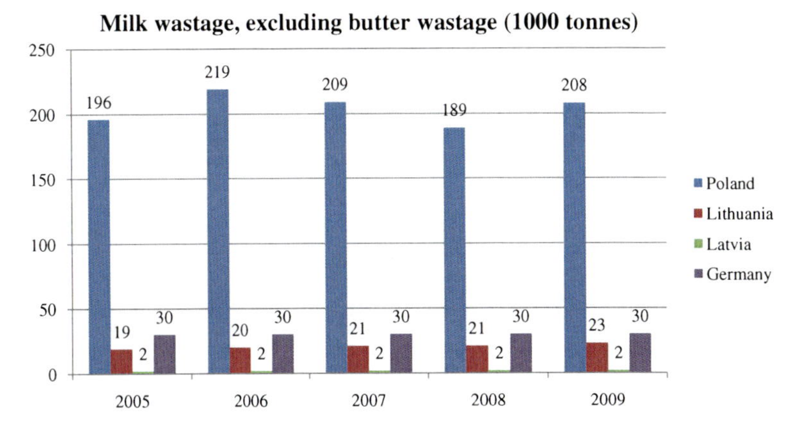

Fig. 7.39 Milk wastage, excluding butter wastage (1000 tonnes) generated in the selected countries in 2005–2009 (FAO 2013a)

In the case of milk wastage, Poland generated the highest amounts of wastage. Moreover, the total amount is more than ten times bigger than in other discussed countries (Fig. 7.39). Poland generated more wastage in the relative terms as well, however, the gap between countries is much smaller (Fig. 7.40).

Secondly, by making a number of rough assumptions based on the discussed above FAO data, the share of food waste for each wastage type could be calculated. The main assumption is that the minimum amount of wastage generated per one hundred tonnes amongst the discussed countries is the unavoidable waste (such as food losses, by-products and food residues) that is accumulated during the stages of the food supply chain. The difference between this minimum value and amounts generated by other countries can be considered as food waste. Thus, zero values in the Table 7.3 indicate that a country generates a minimal amount of that type of waste in comparison to other countries. The calculations show that Poland

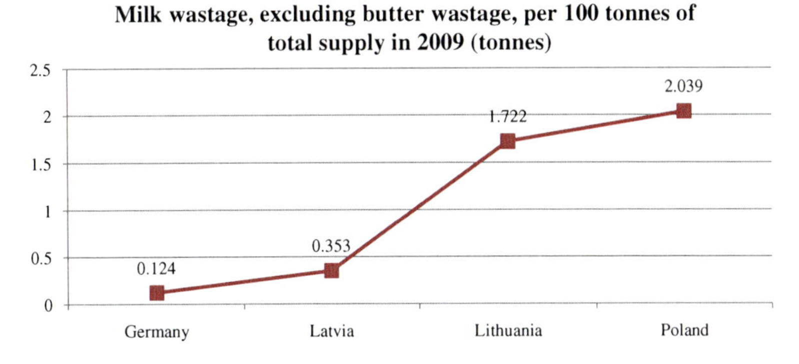

Fig. 7.40 Amount of milk wastage, excluding butter wastage generated by the selected countries per one hundred tonnes, total supply in 2009 (FAO 2013a)

Table 7.3 Amount of food waste generated per one hundred tonnes of total supply, tonnes in 2009 (own calculations based on the FAO Food Supply sheets 2013)

Type of waste	Belarus	Estonia	Germany	Latvia	Lithuania	Poland	Sweden
Fruits—excluding wine:	5.882	N/A	3.24	0	0.343	7.321	2.309
Bananas	0	N/A	9.2419	N/A	N/A	16.971	3.6646
Apples	5.1977	N/A	0	1.631	N/A	5.1129	3.7387
Meat	0.4577	4.6933	0	0.6457	1.4408	0.0955	N/A
Cereals—excluding beer:	1.068	0.158	1.904	0	1.378	2.434	1.223
Barley	0.2463	0.5978	1.8354	0	1.9367	2.732	3.0955
Rye	1.9075	6.4465	2.4177	2.5088	0	3.1846	1.3652
Wheat	1.309	0	3.571	0.175	3.07	3.198	1.936
Pulses	1.6484	N/A	0	N/A	N/A	1.4966	0.3215
Vegetables:	3.849	0	7.252	1.561	2.426	8.222	5.47
Tomatoes	0.2595	1.0145	3.4117	N/A	0	4.9357	0.3704
Onions	0.1509	N/A	6.0833	4.0062	3.4867	4.3493	0
Potatoes	0.5579	1.551	7.8168	0	4.1961	6.0835	5.4065
Eggs	N/A	N/A	0.3477	N/A	N/A	0	1.5312
Milk—excluding butter	N/A	N/A	0	0.229	1.598	1.915	N/A

*N/A—the data are not available

generated the largest amount of fruit, cereal and vegetable waste, whereas Estonia generated the largest amount of meat waste and Germany generated the largest amount of potato waste (Table 7.3).

7.1.2 *Food Waste Generated Based on the FAO Technical Conversion Factors—Extraction Rates*

The amount of food waste could also be calculated based on one of the technical conversion factors used by the Statistics Division of FAO to compile commodity balances and supply/utilization accounts for the countries. The concept of extraction rates relates to processed products only and indicates, in percentage terms, the amount of the processed product concerned obtained from the processing of the parent/originating product, in most cases a primary product. For example, flour from wheat, oil from soybeans, shelled from unshelled almonds, cheese from milk and etc. (Statistics Division of FAO 2000).

The size of an extraction rate defines the amount of food losses, however, it also includes a fraction of food waste that can be calculated by assuming that the maximum value of an extraction rate among the discussed countries is the maximum possible value limited only by currently existing technologies, with no references to any other types of limitations. Therefore, such factors as availability of the technical and financial resources in a country, in order to reach the maximum value of extraction rates, at this stage of calculations are ignored. The difference between maximum and actual values of the extraction rates existing in countries could be considered as food waste. In addition, it is important to note that for some countries the data are unavailable. The results of these calculations show, for example, that Sweden has lowest extraction rates for cereal products i.e. the biggest fraction of avoidable cereal waste (Table 7.4).

The section below includes additional available data regarding the food waste situation in each of the discussed countries.

7.1.3 Belarus

In contrast to other countries discussed in this work, in Belarus, in legal documents, food wastage is defined as organic kitchen waste and that belongs to one of the subtypes of compostable waste. Firstly, such classification was introduced in 2003 in the official guidelines ('Instruction 26'), approved by the Ministry of Housing and Utilities of the Republic of Belarus, for organization of separate collection, storage and transportation of MSW (Resolution No 26 of the Ministry of Housing and Utilities of the Republic of Belarus 2003; Shestakovskiy and Gnedov 2010). Another document that defines food wastage and its subtypes is a 'Classifier of waste types generated in the country' (Resolution No. 63 of the Ministry of Natural Resources and Environmental Protection of the Republic of Belarus 2011). The document is a resolution of the Ministry of Natural Resources and Environment Protection of the Republic of Belarus which came into force in 2011. It categorises several hundred types of waste with their hazard levels into blocks, group and subgroups. According to the classification, food wastage belongs to the block 'Vegetal and Animal waste'. In this block, there are subtypes that could be considered as food waste, for example:

Block I. Vegetal and animal wastes

Group I. Wastes of food and flavouring products

A. Wastes from production of food products
   a. Preserved foods in glass containers with expired dates
   b. Preserved foods in metal containers with expired dates
   c. Preserved foods in plastic package and etc. with expired dates
   d. Other preserved foods with expired dates
   e. …

Table 7.4 Percentage share of food waste based on the extraction rates for the products in the selected countries (own calculations based on the (Statistics Division of FAO 2000))

Type of waste	Belarus	Estonia	Germany	Latvia	Lithuania	Poland	Sweden
Cereals							
Flour of Wheat	1	1	1	4	9	0	4
Bran of Wheat	6	9	4	4	0	10	10
Malt of Barley	0	5	2	7	8	1	8
Flour of Rye	5	9	0	2	7	6	5
Bran of Rye	2	0	7	8	0	10	4
Sugar crops							
The state-of-the-art of the problem of food waste in the Baltic Region countries Beet Sugar	2		0	2	2	2	0
Sugar Refined	4	4	4	0	4	4	4
Molasses	1.7		2.6	1.8	0	1.7	2.7
Beet Pulp, Dry	0		0.4	0	0	0	0.5
Oil-bearing crops							
Oil of Soya Beans			0	1		1	
Cake of Soya Beans			0	4		1	
Oil of Sunflower Seed	11		1	0	0	5	13
Cake of Sunflower Seed	20		9	15	11	7	0
Oil of Rapeseed	6		0	5	3	2	0
Cake of Rapeseed	0		1	15	8	4	5
Oil of Linseed	10		11		10	11	0
Cake of Linseed	5		0		5	3	15
Fruits and berries							
Ferm. Beverages excl. Wine	0	24	36		24	2	
Apple juice Single Streng	10		0	10	10		
Apple juice Concentrated		0	55			40	
Wine	15	25	27	0	16		
Milk							
Cream, Fresh	5	1	4	6	0		7
Butter of Cow Milk	1	0	0.3	1.4	1		0.4
Whole Milk, Condensed	19	19	0	17	21		
		20	0	18	19		

(continued)

Table 7.4 (continued)

Type of waste	Belarus	Estonia	Germany	Latvia	Lithuania	Poland	Sweden
The state-of-the-art of the problem of food waste in the Baltic Region countries Whole Milk, Evaporated							
Dry Whole Cow Milk	13	10	0	12	13		15
Cheese (Whole Cow Milk)	6.3	0	0	3	0		5

B. Wastes from production of flavouring products
   a. Flavouring products with expired dates
   b. ….
C. Waste of food products
   a. Food products with expired dates
   b. Fruits and vegetables that are no longer for consumption (lost their consumption characteristics)
   c. Preserved food in glass and metal containers with expired dates
   d. Food products that are spoiled/damage, contaminated or with no labels
   e. Food products that contain harmful (hazardous) food additives and/or colourings
   f. Food products with higher content of sodium
   g. …

Group III Waste from keeping and processing animals, birds and fish

(Resolution No 63 of the Ministry of Natural Resources and Environmental Protection of the Republic of Belarus 2011).

7.1.3.1 Industries

Available information and statistical data with regard to amounts of food wastage generated and used significantly vary among the sources. Some experts claim that almost all waste from production of food and flavouring products is used in agriculture. According to one of the studies, the amount of vegetal and animal waste generated in 2008 in Belarus was estimated to 3,870 and 4,009 thousand tonnes respectively. The amount includes 1,313 thousand tonnes of wastes from production of food products and 1176.2 thousand tonnes of wastes from production of flavouring products (except lignin and hydrolysed slime), and waste of food products. The authors also noted that 92.6 % and 100 % of these types of waste were used, mostly, for agricultural purposes (Lysuho and Eroshina 2011).

According to other sources, the issue of processing and reusing by-products in the local food industry is still very problematic. Especially in relation to the alcohol, dairy, brewing and fish industries. The alcohol production solely generates more than 1 million tonnes of distillery dregs annually. Usually it is disposed because only 10 % of alcohol production plants have an equipment for processing this type of by-product (Souznoe Veche 2013; Information Analysis Portal of the Union State 2013b).

The production of cheese and quark generates approximately 1.8 million tonnes of thrusting per year. However, in 2008, only 26 % of this amount was processed (Souznoe Veche 2013). According to other sources, this amount did not exceed 3 % (Information Analysis Portal of the Union State 2013b). It is also projected that between the years 2012 and 2015 the production of thrusting will increase up to 2–2.5 million tonnes per year. In comparison, in developed countries 80 % of this by-product is processed and used for production of food products and animal feed (Resolution No. 6 of the Council of Ministers of the Union State 2010).

Only a small share of potato by-products is used for animal feed due to a lack of facilities for processing the feed in most of the potato processing plants. At the same time, the amount of generated by-products is constantly growing. Generation of annual pulp amounts to 20 thousand tonnes (Information Analysis Portal of the Union State 2013b), whereas according to other sources, this value reaches 60–70 thousand tonnes (Souznoe Veche 2013). Production of dried mashed potatoes generates up to 10 thousand tonnes of waste (by-products and residues) and 100–120 thousand tonnes of liquid waste. The biggest part of waste is not used and sent to the sewage system (Souznoe Veche 2013).

Annually, the brewing industry generates about 3 thousand tonnes of malt-sprouts, more than 1.8 thousand tonnes of grain waste, more than 70 thousand tonnes of brewer's wet grains, more than 1 thousand tonnes of protein dregs and more than 3.6 thousand tonnes of residual brewer's yeast as by-products (Souznoe Veche 2013). A lack of technical, normative and legal acts regarding by-products and waste generated as a result of brewing and malt production prevents from producers further use of these resources (Information Analysis Portal of the Union State 2013b).

Fish processing plants generate up to 250 thousand tonnes of fish waste. Waste generated per fish is between 15 and 70 % depending on weight of fish (Resolution No. 6 of the Council of Ministers of the Union State 2010).

7.1.3.2 Retailers

Today, the problem of food waste in retail chains is not out of the main concerns in the country. Firstly, the authorities have to solve the problem of food safety. There is a high number of evidence when small and big retailers sell food products with passed expired dates or products that were spoiled because of wrong storage or handling (Horkova 2013).

7.1.3.3 Municipal Solid Waste

According to the technical code, the document that sets rules for designing and maintenance of landfills for MSW disposal, an approximate average share of food wastage in the total amount of MSW generated annually in a big city equals to 30–38 % (Ministry of Natural Resources and Environmental Protection of the Republic of Belarus 2009), whereas in villages and towns it amounts to 15–25 % (Order No. 14/8a of the Ministry of Natural Resources and Environmental Protection of the Republic of Belarus and the Ministry of Housing and Utilities of the Republic of Belarus 2000).

Thus the share of food wastage in the total amount of MSW generated in 2010 made up to 1016.6 thousand tonnes (27 %).

7.1.4 Estonia

Compared to other EU states, Estonia probably has less problems with the generation of food wastage (Bremere 2011). In 2005 the country generated between 138,308 and 148,048 tonnes of kitchen waste (food residues and food waste) (Zhechkov and Viisimaa 2008).

Furthermore, in comparison to other countries, the share of food wastage generated in the country in total amount of food wastage generated by the EU Member States is very small and equals to only 0.4 % (Fig. 7.1) (BIO Intelligence Service 2010).

In his interview, the manager of the Estonian food bank, Piet Boerefijn noted that in Estonia there are around 200 thousand people who would need food support. Annually, all sectors excluding the agricultural sector throw away around 200 thousand tonnes of food. The general public in Estonia is not very well aware of all the food that goes to waste as well as of consequences for environment. Among the causes of food waste in the country such aspects as a lack of tax incentives on donations, stricter and less clear than in Western Europe food laws, unwillingness of food authorities to give out guidelines, relative low costs of waste disposal and fear of firms to arise scandals and have problems with authorities were named (Boerefijn 2012).

7.1.5 Germany

7.1.5.1 *Waste Generation*

The results of the studies conducted with regard to amount of food waste generated in Germany range between 6 and 20 million tonnes per year. The author of the movie ‘Taste the Waste’, which was considered as one of the important drivers for raising public awareness on the issue of food waste, states that 20 million tonnes of food are thrown away annually (Hallier 2013).

The study conducted by Cofresco 'Wegwerfen von Lebensmitteln/Throwing away food' came out with a number of 6.6 million tonnes of annual food wastage (Cofresco 2013). The study carried out by the University of Stuttgart, published in 2012, reveals a number of approximately 11 million tonnes of food wasted annually. This amount includes all types of food wastage. The experts defined minimum, average and maximum values for different stages of the food chain (Fig. 7.41).

The exact number, 10,970 thousand tonnes, is a sum of mean values of amount of food wastage generated in each stage of the food supply chain (Fig. 7.41). According to the study, the biggest share, 61 %, comes from private households (Fig. 7.42).

The study conducted by the EHI Retail Institute points out that agriculture/processing inclusive bakers, butchers and catering sectors generate 13 million tonnes (as the difference of waste at the distribution and consumer level) (Hallier 2013; Ujhelyi 2013).

According to the EHI Retail Institute GmbH the retail sector generates approximately 310,000 tonnes of food waste (Hallier 2013; Ujhelyi 2013). It does not contain food given to charities. Therefore, the actual amount is higher, and equals to approximately 500 thousand tonnes per year (Kranert et al. 2012). 310 thousand tonnes of food waste include the following categories (Fig. 7.43):

According to the study of the University of Stuttgart, the wholesale sector generates between approximately 43,500 and 87,000 tonnes of waste per year. However, it includes only organic waste and in some cases also plant and flower waste disposed of together with food, therefore, the actual amount of food disposed in this sector is likely to be smaller (Kranert et al. 2012).

The large-scale consumer sector generates between 1,500 and 2,300 thousand tonnes of food wastage per year. The largest share is generated by the catering industry, followed by in-company catering and the accommodation sector (Table 7.5). However, for the researches it was impossible to estimate the share of food waste based on the available literature for individual management types, therefore they used values defined in other works. The approximate percentage share of food waste for this sector ranges between 48.5 and 56 %, thus by applying these values, the amounts of food waste estimates between 727,500–1,115,500 and 840,000–1,288,000 tonnes (Kranert et al. 2012).

The results of the study of the University of Stuttgart regarding the total amount of food wastage generated annually by households coincide with the finding of the Cofresco study. The amount ranges between 5.8 and 7.5 million tonnes, with an average value of 6.7 tonnes. According to the Stuttgart study, the share of food waste equals to 47 %, whereas the Cofresco study points out the value of 59 %, however, it also includes partially avoidable waste (Kranert et al. 2012).

The Federal Ministry of Food Agriculture and Consumer Protection talks about 65 % of food wastage that could be either partially or completely avoided (BMELV 2013b). Thus, by applying these values to the given range of the amount of food

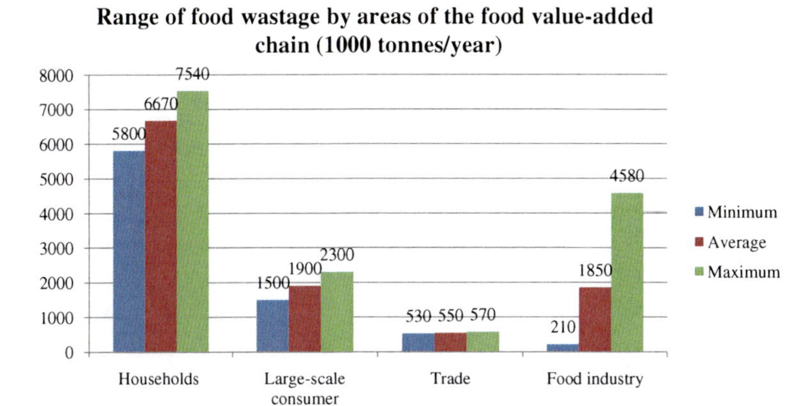

Fig. 7.41 Range of food wastage in Germany by areas of the food value-added chain, 1000 tonnes/annum (Kranert et al. 2012)

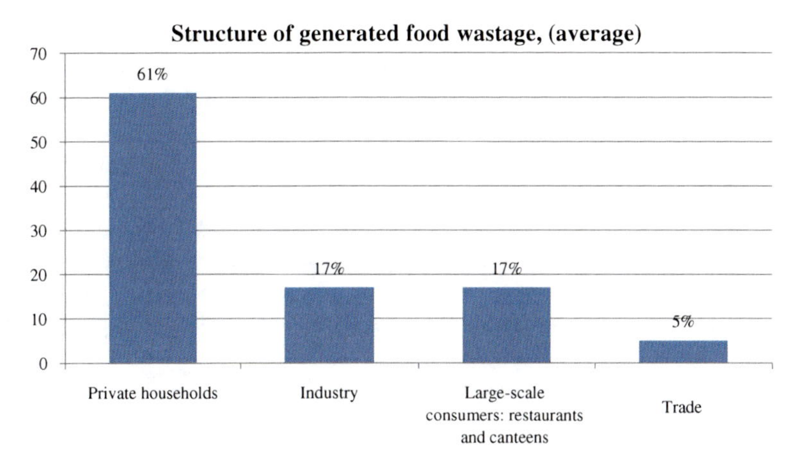

Fig. 7.42 Percentage ratio of food wastage generated by different sectors (Kranert et al. 2012)

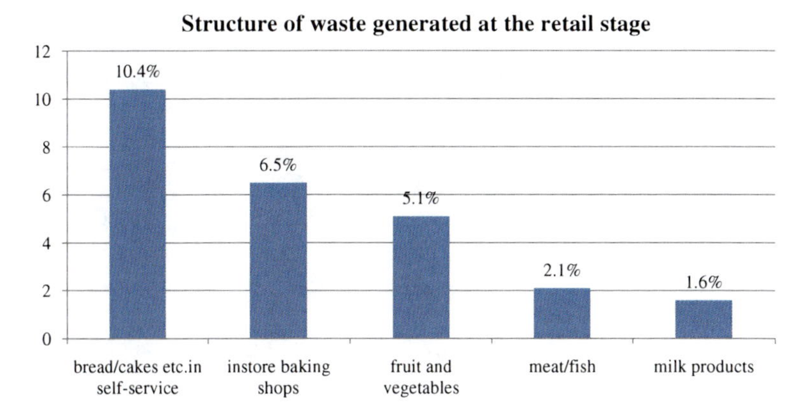

Fig. 7.43 Structure of waste generated at the retail stage (Hallier 2013; Ujhelyi 2013)

Table 7.5 Amount of food wastage generated by each sector (Kranert et al. 2012)

Sector	Food wastage generated, tonnes per year
Catering industry	837,000–1,015,000
In-company catering	147,000–402,000
Accommodation sector	186,000
Retirement and nursing homes	93,000–145,000
Schools	75,000–87,000
Hospitals	65,000
Universities, day-care centres for children, prisons, German armed forces (Bundeswehr)	less than 41,000

wastage, the share of food waste estimates between 2.726 and 4.875 million tonnes (Table 7.6).

According to the Federal Ministry of Food Agriculture and Consumer Protection (BMELV), the most common foods thrown away are fruit and vegetables, which make up to 44 % of all food waste in private households (BMELV 2013b). According to the Stuttgart study, the largest share of avoidable and partly avoidable household food wastage comprise of vegetables (26 %), fruits (18 %), pasta and bakery products (15 %) (Fig. 7.44).

However, the results of the research project conducted by the Institute of Sustainable Nutrition and Food Production (iSuN) of the Münster University of Applied Sciences and the Consumer Center of North Rhine-Westphalia show that bread is wasted on a more frequent basis than fruit and vegetables, and leftover foods. Meat, dairy and cereal are thrown away rarely. It also suggests that food which comes without an expiration date is discarded more often than products with a set best-before or use by date (University of Applied Sciences Münster & Institute for Sustainable Nutrition and Food Production—ISuN 2012).

7.1.5.2 Causes

The variety of studies came out with such causes of food waste as overproduction that results from a gap between estimated sales and actual sales (Hallier 2013), bad planning due to unstable demand for food, production losses and faulty batches

Table 7.6 Amount of food waste generated by households for the given range of percentage share (own calculations based on (Kranert et al. 2012; BMELV 2013d)

%	5.8 million tonnes	7.5 million tonnes
47	2.726	3.525
59	3.422	4.425
65	3.77	4.875

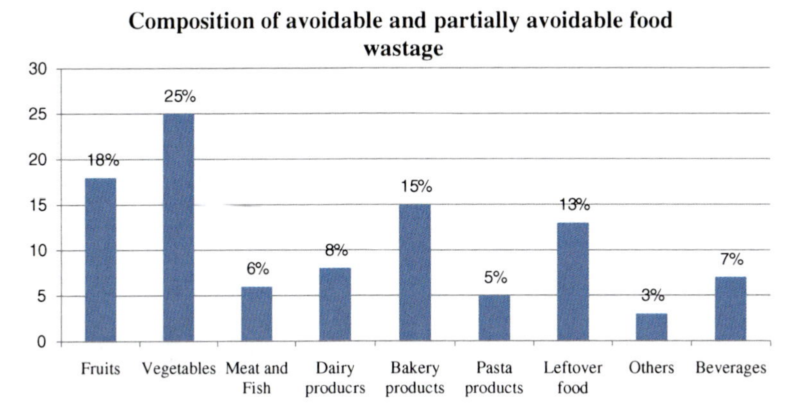

Fig. 7.44 Composition of avoidable and partly avoidable households food wastage by product group, percentage by weight (Kranert et al. 2012)

caused by technical problems in the food industry (Kranert et al. 2012) (Fig. 7.45), as well as portions, storage, regulations, best before date, estrangement from food, full shelves till closing time, imperfection and etc. (Braun 2012).

However, most representatives of the food industry, surveyed during the Stuttgart study, noted that it is nearly impossible to avoid waste of food at the manufacturing stage of the value-added chain (Kranert et al. 2012).

The iSuN and the Consumer Center of North Rhine-Westphalia study identified the following main causes of disposal of different product groups:

- Vegetables: commission regulations, product specifications, and standardised packaging;
- Bread and baked goods: the limited freshness of the goods that conflict with the consumers' expectations to find the goods available until late in the evening;

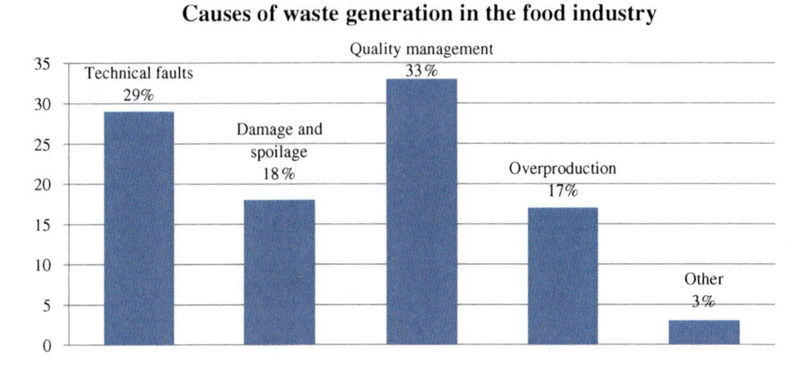

Fig. 7.45 Reasons for waste generation in the food industry (Kranert et al. 2013)

- Milk and dairy: technical issues (for example, losses through defects of the machinery) or other deficiencies (such as losses due to cleaning of the machines or a change of flavour) and expiration dates;
- Meat and sausage products: any deviations from product specifications regarding quality, optical characteristics, texture and temperature automatically lead to its discard, because foods from this product group hold high health risks

(University of Applied Sciences Münster & Institute for Sustainable Nutrition and Food Production—ISuN 2012).

Trouble with the planning of meals for bigger groups of people was named as one of the causes of household food waste (University of Applied Sciences Münster & Institute for Sustainable Nutrition and Food Production—ISuN 2012). Cofresco also noted that most consumers strongly underestimate the amount of food they throw away. Many are convinced that they do not waste any food at all (Cofresco 2013). The researchers also came to the conclusion that the expiration date is not misunderstood, but simply used as an excuse for the discarding of products that consumers dislike.

On the other hand, they stated that a better knowledge about the correct storage of food correlated with a better understanding of best-by and use-by dates. In addition it was pointed out that the kind and amount of food waste depends among others factors on the available products, the living environment, the needs, abilities, resources and the personal level of commitment (University of Applied Sciences Münster & Institute for Sustainable Nutrition and Food Production—ISuN 2012).

7.1.6 Latvia

According to the study, in the years 2004–2006, the amount of food wastage generated by the local food industry, which is the second largest industrial sector in Latvia (Vorne 2012), ranged from 3664.4 to 63732.73 tonnes depending on the region in the country (Dzene 2009). In the year 2006 the largest share came from the dairy production (Fig. 7.46).

According to the study from the Barilla Center for Food & Nutrition, food wastage per capita, which is generated during industrial processing, distribution, and final consumption amounts to 94 kg per year (Barilla Center for Food and Nutrition 2012) that in total makes up approximately 190,256 tonnes per year. This number differs from that presented by BIO Intelligence Service (Table 7.1). In 2006, the collected amounts of biological kitchen waste and waste from marketplaces were 50 and 546 tonnes, respectively (Dzene 2009).

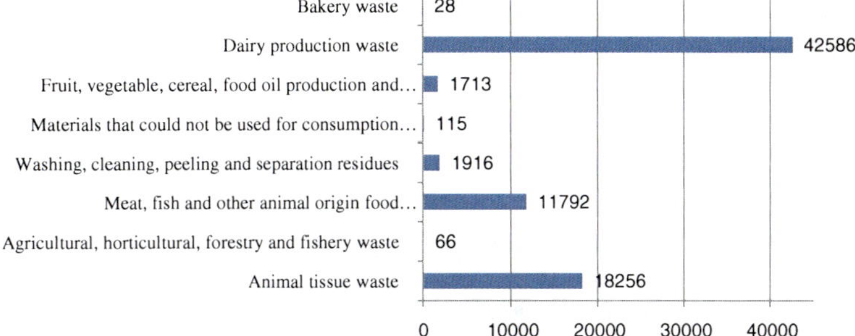

Fig. 7.46 Amount of food wastage from food industry in 2006 in tonnes (Dzene 2009)

7.1.7 Lithuania

According to the results of the research conducted by the Lithuanian Food bank, 79 % of food wastage generated in the food supply and production sector is edible and safe for human consumption (Tylaite and Bastys 2013). The study of the Barilla Center for Food and Nutrition states that per capita food wastage generated during industrial processing, distribution, and final consumption stages amounts to 171 kg per year (Barilla Center for Food and Nutrition 2012) that in total is approximately 508,212 tonnes per year.

According to Juškaitė-Norbutienė et al. 2013, animal by-products generated by meat, dairy and fish industries in 2005, amounted to approximately 286 thousand tonnes, where the biggest share belongs to the dairy industry (Fig. 7.47). It mainly includes whey, reject and end-of-life products (Juškaitė-Norbutienė et al. 2013).

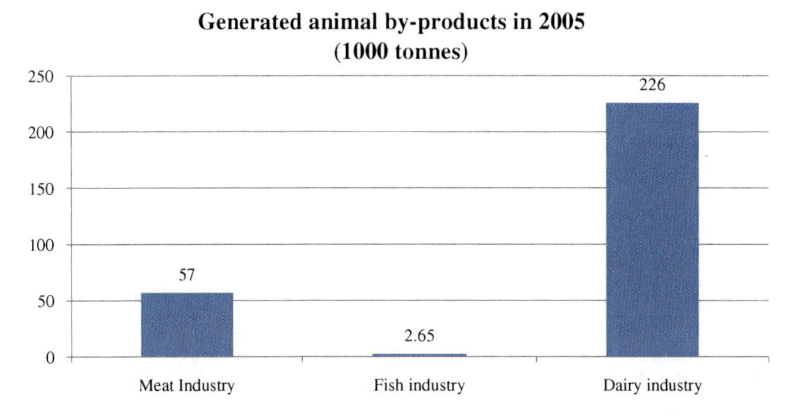

Fig. 7.47 Amount of animal by-products waste generated by the industries in 2005, 1000 tonnes (Juškaitė-Norbutienė et al. 2013)

In 2010, the total amount of biodegradable waste and by-products generated by these industries exceeded 400 thousand tonnes, with the largest share still coming from the dairy industry (Juškaitė-Norbutienė et al. 2013).

According to another study, food wastage generated by the food industry (meat, diary, mill and bakery) between the years 2004 and 2007 amounted to approximately 313 thousand tonnes per year (Fig. 7.48).

The waste from the bakery industry is characterised by two types—expiry-date products and manufacturing waste. The latter is estimated to 0.1–4.3 % from the production volume, depending on the product nature, technology, sector, season, and human factor, in the case of small companies with low rates of mechanization. According to the manufacturers, in large companies, it is virtually impossible to reduce the rate of waste generation due to the specifics of production (Juškaitė-Norbutienė et al. 2013).

In the case of the food industry one of possible causes of food waste is the annual EU production quotas that limit milk and sugar production and prohibit the export of overproduced quantities. In certain situations, when there are no other possibilities for alternative use of these products, these amounts automatically become waste.

During the study conducted by the Lithuanian Food Bank, respondents indicated the following causes of food waste:

- Product expires in the warehouse;
- Seasonal products, seasonal packaging;
- Wrong label on a product;
- Mistakes in planning the demand;
- Mistakes in planning the sales volume;
- During the selling process, when buyer changes his/her mind and leave the short term product anywhere in the shop

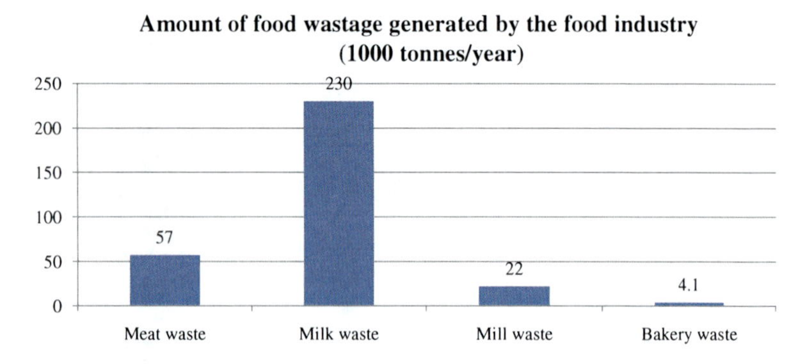

Fig. 7.48 Amount of food wastage generated by the food industry in 2004–2007 (Havukainen et al. 2012)

(Tylaite and Bastys 2013).

In addition, it was found that companies do not have enough information and knowledge about possible ways of managing food waste e.g. eligibility for donation, ways of starting the process, lack of information about (charitable) organisations that work with food waste, tax regulations on donating food (e.g. VAT, tax cuts), regulations on food safety for food waste donations. The respondents also express a willingness to get some kind of benefits (e.g. lower costs, publicity) from this type of activity.

The study showed that there is no online information about the food waste reduction and recovery in Lithuania, except the Food bank webpage and blog. Furthermore, there are no educational programs created for this purpose (Tylaite and Bastys 2013).

7.1.8 Poland

In the country, where several million people live in deep poverty (Gosiewska 2013), it was revealed that a variety of food products with close to expiry or past dates such as vegetables, fruit, dairy, sweets, eggs, even luxury goods as expensive cheeses and chocolates can be found in waste bins (Ciobanu 2013).

According to BIO Intelligence Service (2010), the largest share of food wastage comes from the manufacturing sector (BIO Intelligence Service 2010). In 2012, the federation of Polish Food Banks collected 7,427 tonnes of food through the cooperation with the food industry (Ujhelyi 2013; Gosiewska 2013).

Food residues and by-products generated by the fruit and vegetable industry comprise mainly of fruit and vegetable pomace and peeling. However, due to low demand, these resources are not entirely used as animal feed (Malinska 2004).

As a result of poor orchard management overall food waste in traditional fruit storage may account for more than 20 %, through rot and mould (Foresight 2010).

According to the Polish classification, food wastage includes wastes from agriculture, horticulture, aquaculture, forestry, hunting and fishing, food preparation and processing (Polish Council of Ministers 2010). The Table 7.7 indicates types and amounts of food wastage generated by industries in 2004, 2006 and 2008. It is important to note that a very large decrease in the mass of waste from the sugar industry in 2006 is a result of reduced sugar production due to the EU quotas (Polish Council of Ministers 2010).

According to the Barilla Center for Food and Nutrition, per capita food wastage generated during the industrial processing, distribution, and final consumption stages is 235 kg per year, which in total amounts to approximately 9,055,255 thousand tonnes per year (Barilla Center for Food and Nutrition 2012).

Withregard to the causes of food waste, the results of available researches indicated similar causes of food waste generated by industries and households in Poland that were identified in other countries.

Table 7.7 Amount of food wastage generated in 2004, 2006 and 2008, broken down by types (Polish Council of Ministers 2010)

Waste name	Amount of non-municipal food wastage, Mg thousand/year		
	2004	2006	2008
Animal-tissue waste	31.2	21.0	3.1
Plant-tissue waste	21.9	40.8	53.2
Wastes from agriculture, horticulture, hydroponic cultivation, forestry, hunting and fishing	265.4	296.1	444.1
Animal-tissue waste	265.8	653.5	523.1
Materials unsuitable for consumption or processing	22.4	30.1	35.4
Waste from production of fish flour	0.0	0.0	0.0
Other wastes from the preparation and processing of foods of animal origin not otherwise specified	323.9	752.8	616.0
Pomace, settlings and other waste from vegetable products processing (excluding waste from plant feed production)	211.1	254.8	379.6
Waste from plant feed production	0.0	1.2	0.0
Waste from the preparation and processing of products and semi-luxury food and drinks and waste of plant origin (excluding wastes from the production of alcoholic and non-alcoholic beverages (except coffee, tea and cocoa))	305.0	647.9	753.9
Beet pulp	4228.6	3221.7	1707.3
Waste from sugar industry	4228.8	3223.8	1707.3
Materials unsuitable for consumption or processing	63.9	99.6	6.0
Sludge from on-site effluent treatment	21.7	22.4	17.2
Waste whey	807.2	963.6	933.6
Wastes from the dairy products industry	892.8	1,085.6	956.8
Materials unsuitable for consumption or processing	6.6	10.1	9.7
Wastes from the baking and confectionery industry	6.6	10.1	9.7
Wastes from washing, cleaning and mechanical reduction of raw materials	15.7	28.5	13.7
Materials unsuitable for consumption or processing	3.6	0.9	0.9
Total	13448.6	14144.2	11240.6

7.1.8.1 Industries

During the survey conducted by the Polish Food Bank, companies named such causes of producing food waste as process losses during the primary and secondary processing, discard of products during the product evaluation (e.g. quality control, standard recipes), damage during transport e.g. spoilage and poor handling in wet market (Gosiewska 2013). In addition, there are lack of internet platform for on-line communication among NGOs, food producers, distributors, restaurants and food farms that would provide an efficient system of collecting short-date food; lack of knowledge about legal aspects regarding food donation, as well as lack of financial incentives for food donating companies (Gosiewska 2013).

7.1.8.2 Households

According to the Food Bank, among the main causes of household food waste are:

- not using food in time
- too large portions of meals
- too large shopping
- improper storage
- poor quality of the product
- lack of ideas to use the components for a variety of dishes

(Gosiewska 2013).

The survey, conducted by Public Opinion Research Center, Poland (CBOS) in 2005 on 'culinary tastes, eating habits and consumer behaviour Poles', found that wasteful behaviour clearly linked to the level of an households income and assessment of their own material status (Ujhelyi 2013; Gosiewska 2013).

It is important to note that in contrary to other countries, where a research shows that usually consumers are not aware of amount and environmental consequences of food that is thrown away, in Poland, according to the study of the Food Bank, 30 % of respondents admit to have food waste at home (Ujhelyi 2013; Gosiewska 2013). Furthermore, 85 % of population is aware that food waste is waste of money, 71 % believe that it is harmful for the environment, 66 % of the respondents stated that wasted food has a significant impact on food prices and 83 % of respondents admit that wasting food is a social problem (Ujhelyi 2013; Gosiewska 2013).

7.1.9 Sweden

The first research findings regarding food waste in food service institutions in the country were published in 1979 and had the following structure (Fig. 7.49):

However, this study was only based on few institutions. The next study was conducted in 1985 and covered the topic of household waste in 90 households. The results showed that 16 % of the potatoes, 3 % of the bread and 17 % of the meat were wasted (Engström and Carlsson-Kanyama 2004).

In 2001, the results of the study conducted in two schools and in two restaurants in Stockholm showed that on average, 20 % of the food delivered was wasted. About 96 % of this amount was waste, generated as a result of improper storage. It was also comprised of food that was left on serving dishes, in canteens and bowls, food that was prepared but never served and plate waste (Engström and Carlsson-Kanyama 2004).

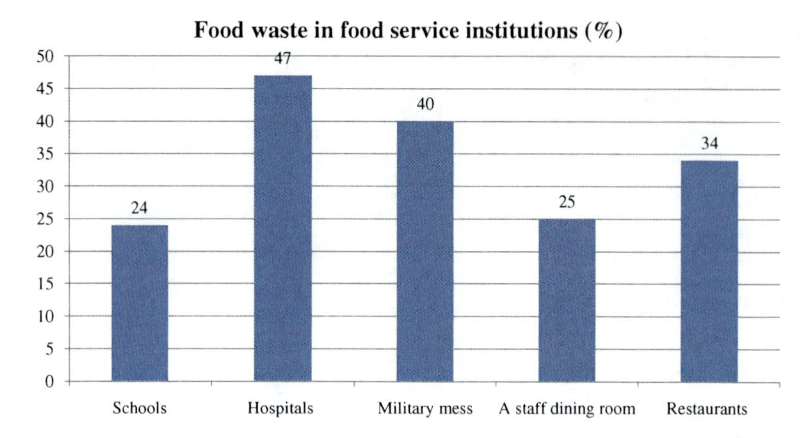

Fig. 7.49 Structure of food waste in the food service institutions in 1979 (Engström and Carlsson-Kanyama 2004)

The statistical data regarding amounts of food wastage and a share of food waste generated between the years 2010 and 2012 vary in different sources.

According to Swedish Methodology for Environmental Data (SMED), that performed a study for the Environmental Protection Agency of Sweden (Naturvårdsverket), in 2010, the country generated over a million tonnes of food wastage, along the whole food supply chain. 67 % of this amount is derived from household consumption (of which 35 % is represented by food waste that correspond to 259,000 tonnes). In the same year, the Swedish food industry generated about 17 % of total food wastage (Table 7.8) (Barilla Center for Food and Nutrition 2012).

7.1.9.1 Households

In 2004, according to the Swedish Environmental Protection Agency, 40 % of household waste consisted food wastage (Swedish Environmental Protection Agency 2005). The work of Westregård combines the results of a number of studies conducted in different years with regard to amounts of food waste in the country (Table 7.9). According to that sources between years 2004 and 2012, the total amount of food wastage ranged between 910 and 848 thousand tonnes. The share of food waste varied between 35 and 57 % (Westregård 2012).

Currently, the total amount of household food wastage is estimated between 910 and 920 thousand tonnes per year (Modin 2011; Eco-Innovation Observatory 2013). However, experts differ in their opinions with regard to the share of food waste.

Table 7.8 Food wastage in Sweden in 2010 SMED 2011 in (Barilla Center for Food and Nutrition 2012; Marthinsen et al. 2012)

Sector	Generated amounts (tonnes)	Percentage of total generated amount (%)
Farming	unknown	Unknown
Food Industry	171,000	17
Grocery stores	39,000	4
Restaurants	99,000	10
School canteens	26,000	3
Household:	674,000	67
Included unavoidable waste (e.g. food residues)	435,000 (65 %)	–
Included food waste	259,000 (35 %)	–
Total	1010,000	–

Table 7.9 Total amount of food wastage and a share of food waste generated according to different studies (Westregård 2012)

Year	Food wastage (tonnes)	Household food waste (tonnes)	Household food waste (%)
2004	910,000	–	–
2008	–	–	57
2011	674,000	239,000	35
2012	848,000	297,000	35

For example, according to Modin 2011, the share of the food waste is likely to be slightly above 50 % of the total amount that corresponds to 455,000 tonnes (Modin 2011).

According to the Saraiva Schott et al. 2013, the average fraction of food waste among households reaches 34 % or 272,000 tonnes on an annual bases (Saraiva Schott et al. 2013).

The Swedish Waste Management and Recycling association claims that between 10 and 20 % of all purchased food is discarded (Avfall Sverige 2012).

In 2011, the quantity of collected household food waste amounted to 275,000 tonnes (Avfall Sverige 2012).

According to the National Food Agency, an average annual amount of perfectly edible food thrown away in total amounts to 238,900 tonnes per year (Swedish National Food Agency 2013). Fruits and vegetables, bread and dairy products are products that thrown mostly (Modin 2011; Stenmarck et al. 2011).

7.1.9.2 Retail Sector

In 2008, the retail sector generated 83,500 tonnes of food wastage where the biggest share was food waste (Stenmarck et al. 2011). According to the results of the study conducted in cooperation with one of the leading Swedish retail companies between 2008 and 2009, retail waste of horticultural products amounted between 0.4 and 6.3 % of store supplies for different horticultural products (Gustavssona and Stagec 2011). The analysis of food waste in small and large stores shows that products that are sensitive to rough handling, such as broccoli, strawberries and cauliflower, registered the highest shares of waste, regardless of the size of a store (Table 7.10) (Gustavssona and Stagec 2011).

According to the Allwin company which distributes food to people in social need, an average amount of perfectly good food thrown by retail stores equals to 60 thousand tonnes per month (Eisner 2012). This value significantly exceeds the statistical data pointed in other sources.

The results of the study conducted at six Swedish retail stores (the stores belong to the low price sector of the Swedish retail food market) regarding flows of fruits and vegetables, indicated that average fruit and vegetable wastage amounted to 4.3 % by mass in relation to quantity delivered. 3.01 % of this is pre-store waste, i.e. fresh fruits and vegetables (FFV) items rejected by a store at delivery, due to non-compliance with quality requirements. 0.99 % is FFV waste occurring after purchase from a supplier and mostly depends on exceeded best-before dates or product deterioration. 0.3 % is unrecorded in-store waste, i.e. underestimated mass during recording unpackaged waste as well as unrecorded wasted items (Eriksson et al. 2012).

The Swedish Environmental Protection Agency states the following values for food wastage:

- Retail wholesale sector—super markets: 74,000 tonnes
- Retail wholesale sector—smaller shops: 9,500 tonnes

(Stenmarck et al. 2011)

7.1.9.3 Hospitality Sector

In 2004, the findings of a study that analysed four different restaurants and school kitchens indicated that around 16 % of the purchased food was cooked but not eaten and proportion of plate waste equalled to 10 % (Schneider 2008).

Between the years 2002 and 2009, a number of studies were conducted in regard with food waste at a school kitchen. The results demonstrated that a share of food waste per plate/portion of served food ranged between 7.5 and 9.6 %. The total share of food waste that included a plate waste, waste in storage, preparation and serving was estimated to 18 % (Marthinsen et al. 2012). Thus, based on an average portion weight of 0.333 kg and a number of portions of food totally served in Swedish schools restaurants per school day (1.4 million), the total amount of food

Table 7.10 Annual average proportion of wasted store supplies for small and larger stores (Gustavssona and Stagec 2011)

Small stores		Large stores	
Celery	7.43 %	Broccoli	7.38 %
Strawberries	6.47 %	Kiwi fruit	4.54 %
Cauliflower	5.71 %	Cauliflower	3.81 %
Swedes	5.50 %	Swedes	3.24 %
Broccoli	5.12 %	Parsnips	2.88 %
Parsnips	3.92 %	Celery	2.42 %
Iceberg lettuce	3.25 %	Strawberries	2.31 %
Kiwi fruit	3.15 %	Pears	2.22 %
Pears	2.99 %	Tomatoes	1.64 %
Tomatoes	2.87 %	Leeks	1.63 %
Leeks	2.64 %	Carrots	1.49 %
Apples	1.41 %	Apples	0.88 %
Cucumbers	1.34 %	Iceberg lettuce	0.81 %
Carrots	0.96 %	Cabbage	0.61 %
Cabbage	0.76 %	Cucumbers	0.61 %
Onions	0.68 %	Onions	0.21 %

Small stores		*Large stores*	
Celery	7.43 %	Broccoli	7.38 %
Strawberries	6.47 %	Kiwi fruit	4.54 %
Cauliflower	5.71 %	Cauliflower	3.81 %
Swedes	5.50 %	Swedes	3.24 %
Broccoli	5.12 %	Parsnips	2.88 %
Parsnips	3.92 %	Celery	2.42 %
Iceberg lettuce	3.25 %	Strawberries	2.31 %
Kiwi fruit	3.15 %	Pears	2.22 %
Pears	2.99 %	Tomatoes	1.64 %
Tomatoes	2.87 %	Leeks	1.63 %
Leeks	2.64 %	Carrots	1.49 %
Apples	1.41 %	Apples	0.88 %
Cucumbers	1.34 %	Iceberg lettuce	0.81 %
Carrots	0.96 %	Cabbage	0.61 %
Cabbage	0.76 %	Cucumbers	0.61 %
Onions	0.68 %	Onions	0.21 %

waste per school day reaches approximately 84 tonnes that correspond to about 15 thousand tonnes per year. According to more recent data from the Swedish Environmental Protection Agency, total food waste from Swedish schools amounts to approximately 10,000–30,000 tonnes per year (Modin 2011).

Estimates suggest that 20 % of food is wasted in food service institutions. According to the Environmental Protection Agency of Sweden, the total amount of food wastage from the service sector is estimated to 377,000 tonnes (Foresight 2011).

The study of food waste, conducted in the frame of the Nordic study, claims that the total amount of food wastage generated in the hospitality sector is equal to 260,000 tonnes per year, out of which 174,000 tonnes (approx. 67 %) including liquid waste, is food waste (Sundt 2012). Whereas, according to the Swedish Environmental Protection Agency, the total amount of food wastage from the hospitality sector is estimated to approximately 300,000 tonnes (Marthinsen et al. 2012).

Today, restaurants, food stores and schools in Sweden have the possibility to sort their food wastage. However, an analysis shows that a significant fraction of this type of waste still ends up in the mixed municipal waste. An analysis of waste composition, performed by the City of Stockholm, suggested that 39 % of MSW consists of food wastage (Millers-Dalsjö and Lundborg 2012). Another study pointed out that the share of food wastage in municipal waste that comes from restaurants equals to 20 %, from food stores to 40 % and from schools to 13 %. Thus, the amount of food wastage generated by restaurants is estimated to 127,000 tonnes, by food stores to 67,000 tonnes, and by schools to 30,000 tonnes. In addition, the authors pointed out that the share of food waste in these amounts equals to 62 %, 91 % and 52 % respectively (Stare et al. 2013).

7.1.9.4 Causes

According to the results of the interviews, conducted in the framework of the Nordic study, the main cause for food waste generation from retail shops and the wholesale sector is that the food turns 'un-saleable', e.g. food with the passed best before/expiry date or lack of freshness for perishables. The respondents also mentioned such causes as improper storage/handling of food products, challenges in ordering because of difficulties to predict consumers' demand, return of unsold products to food producers with no cost for a shop owner; break-down of products due to wrong type of mechanic handling. In addition, it was noted that often additional 7 % to the expected sales are produced in order to meet customers' expectations of full shelves throughout the opening hours of shops (Stenmarck et al. 2011).

7.2 Food Waste Treatment

According to the food waste recovery hierarchy, presented by the U.S. Environmental Protection Agency (Fig. 2.9), the most preferable stages of food waste treatment are 'source reduction' and 'feed of hungry people'. Currently, countries apply a variety of methods that fall under these types of treatment. It is important to note that none of these methods alone is sufficient to achieve the reduction targets.

Implementation of any treatment method requires engagement and cooperation of different parties together with realisation of other supplementary activities.

One of the first steps in the process of problem solving is recognition of the fact that a problem is actually exist. In the case of the problem of food waste, it could be done through public awareness and education campaigns. The next steps are more practical such as food donation and a change of behaviour, followed up by biological treatment and incineration/disposal, which are considered least preferable methods.

Today, each of the discussed countries implements some of these methods that appear to be less or more efficient to the solution of the problem. Due to the need to provide a better understanding of the situation in each country, a description of the situation in individual countries follows.

7.2.1 Belarus

Since the problem of food waste is not well investigated, at least in comparison to other countries discussed here, it is possible to assume that food waste is treated in the same way as the other fractions of bio-waste. An analysis of the treatment practices in the country shows that the issue is more developed in the legal area, i.e. more theoretically than practically. There is a variety of documents, regulations and instruction that prescribe, mostly, bureaucratic activities in regard with the food wastage treatment. It is also important to note that all these regulations are focused on the methods from the lower part of the food waste hierarchy and mostly do not cover food wastage generated by households.

Analysing the current 'waste situation' in the country it could be said that the problem of food waste, particularly of household food waste, is not one of the foremost issues of waste management in Belarus. There are a number of more urgent problems that should be solved first, for example, problems in the municipal solid waste collection system, such as a lack of unified standards for waste containers and their uncoordinated placement, caused by lack of approved schemes that consider residential density (Mihalap and Plepis 2012). Currently, about 90.4 % of municipal waste goes to 170 large and 3,699 mini sites. Another problem is the waste tariffs for households. The size is calculated based on the standard rate of waste generation and not on actual generated and disposed amounts, and, therefore, it does not include all associated costs. The difference is covered by waste tariffs, paid by companies (Mihalap and Plepis 2012).

Currently, the Ministry of Housing and Utilities of the Republic of Belarus has brought up for the public discussion a draft of the Technical Code of Practice 'Environmental protection and use of natural resources. Waste. Guidelines/Rules for treatment of municipal solid waste'. According to the document, separate waste collection must be applied before any further waste treatment (Mihalevich 2013). However, at the same time it states that separate collection of food wastage from apartment buildings might be implemented only if there are composting facilities or other possibilities for alternative uses of this type of waste in the residential area (Ministry of Natural Resources and Environmental Protection of the Republic of Belarus 2013a).

According to another document 'Instruction 26', when it is possible, compostable waste should be composted by producers of municipal solid waste, including households, on their land parcels, personal subsidiary plots and etc. However, the document does not set any rules, guides or regulations for such composting (Shestakovskiy and Gnedov 2010).

According to the legislation, all facilities that use waste are required to be officially registered. Currently, these facilities use only 44 % of total waste generated. Only 0.3 % of all registered facilities use food wastage, generated during food production (Lysuho and Eroshina 2011).

Another document that is used across the country is the instruction, 'Organisation, collection, storage, recording, transportation and disposal of food wastage'. The document is intended for use in the hospitality and retail sectors, as well as in educational and health institutions. According to the instruction, all organizations should register their food wastage. All products with passed dates, without labels, spoiled and etc. (i.e. food waste) must be registered separately, with indication of a reason for discard. According to the document collected food wastage, firstly, must be denaturised, and then landfilled together with other municipal solid waste (Ministry of Housing and Utilities of the Republic of Belarus 2013).

The manufacturing sector is almost the only sector where measures aimed at reduction of food waste have practical implementation. In order to reduce a food waste fraction in the amount of generated by-products, the Ministry of Agriculture and Food of the Republic of Belarus together with the Ministry of Agriculture of the Russian Federation have initiated a research programme that focused on an increase of effectiveness of food production by processing generated waste with advanced technologies and equipment. The main objectives of the programme were to develop 18 technologies and 20 types of equipment for re-equipment of the food industry. It was planned that these developments would contribute to introduction to environmentally friendly and resource-saving processes and make possible the production of valuable feed and food products from generated waste. The practical outcomes of the programme are implementation of 8 pilot projects, including 2 in the alcohol industry, 3 in the dairy industry, and 1 in the meat, beer and potato processing industries. Among others, the developed equipment is used for processing alcohol stillage and whey (Information Analysis Portal of the Union State 2013a).

7.2.2 *Biological Treatment in Belarus*

It is important to note that the country is quite active in developing its potential in the area of biological waste treatment. According to the Head of Representative Office of German Economy in the Republic of Belarus, Vladimir Augustinsky, the country has the highest potential of producing biogas from waste that comes from animal, poultry and food industries (Kalinovskaya 2013).

In one of the interviews, the Deputy Minister of the Ministry of Housing and Utilities of the Republic of Belarus, Anatoly Shagun, noted that the country has started treatment of a biodegradable fraction of municipal waste. For these purposes in one of the cities, separate collection of food wastage from the hospitality and industry sectors has been set (Trofimovich 2013).

There is also an example of international cooperation, a joint European Union and United Nations Development Programme Project 'Support to the development of a comprehensive framework for international environmental cooperation in the Republic of Belarus'. One of the on-going initiatives is distribution of bio-composters for home composting. In 2012, in the frame of the project, facilities and equipment, necessary for separate collection of MSW were transferred to the local office of the Housing and Utilities (Joint European Union and United Nations Development Programme Project 2013).

7.2.3 Estonia

7.2.3.1 Public Awareness Campaign/Education

According to the study 'Improving Waste Prevention Policy in the Baltic States. Assessment and Recommendations' (Bremere 2011), for promotion and stimulation of sustainable distribution and consumption, such instruments as awareness raising campaigns, education at schools, technical training courses, green public procurement, labelling and certification are applied in the country.

7.2.3.2 Food Donation/Food Banks in Estonia

First Estonian Food Bank was opened in 2010. Today, there are 11 food banks in the country. Among food donors are supermarkets, wholesale firms, producers, farmers and private donors. The organisation also gets support from the EU food programme (PEAD) (Boerefijn 2012). According to the Institute of Development Studies, in 2011, the Estonian Food Bank served more than 95 thousand people (Gentilini 2013). In the interview, Piet Boerefijn, the manager of the Estonian Food Bank, stated that in 2012, the organisation distributed around 900 tonnes of food, 50 % of which came from the PEAD.

Among the reasons for not transferring potential food waste to the Food Bank, Piet Boerefijn named the following:

- food banks are quite new in the country;
- some firms have to pay value added tax or income tax if they donate food to the Food Bank, therefore, disposal is much easier and cheaper method;

- some firms are afraid of reputation or fraud if they donate food to the Food Bank;
- the Estonian Food Bank has very little financial resources, so it is difficult to build up a professional warehouse- and transport system, 95 % of the work is done by volunteers;
- Estonian public does not have sufficient knowledge about amount of food waste generated and its environmental consequences; as a result, there is not much pressure on firms not to throw food away.

Piet Boerefijn also added that the current legislation makes it difficult to firms to donate food, and, therefore, the Food Bank is battling for last 4 years with food authorities, tax authorities and ministries to make changes and simplify the food donation process. He noted that the only reason the government takes some measures is due to EU requirements. This statement is also 'supported' by the national environmental strategy until 2030 (Estonian Ministry of the Environment 2007) that does not discuss any issue related to the problem of food waste.

7.2.3.3 Separate Collection

For municipalities the collection of food wastage is voluntary (BiPRO 2012a). In 2007, the city of Tallinn has implemented the separate collection of biodegradable kitchen waste. The regulation was adopted because of the Estonian targets regarding biodegradable municipal waste (BMW). According to the regulation, buildings with more than five apartments must have a separate container, as well as organisations, producing more than 25 kg of BMW per week. Experts also pointed out that, since there has not been any tradition in this type of separate collection, people need time to get used to the system. Therefore, in the first few months the results of implementation were poor (Zhechkov and Viisimaa 2008). Currently, only ca. 15 % of total kitchen waste is source-separated and goes to composting facilities (BiPRO 2012a).

7.2.3.4 Biological Treatment

There are a number of biodegradable waste treatment methods implemented in the country that potentially might be used for food waste treatment as well. Biodegradable kitchen and canteen waste mixed with garden and park waste, municipal wood waste and some paper and other cellulose-based waste are sent to organic recycling (Estonian Environment Information Centre 2010).

In 2011, plans were to build an anaerobic digestion plant for sludge and biowaste, and 2 composting plants. In the same year, 4 MBT plants with a total capacity of 300,000 tonnes and 2 composting plants were already operated (Moora 2011).

7.2.3.5 Additional Activities/Initiatives in Estonia

Experts have named economic arguments as the main reason for prevention of food waste in different institutions that strive to decrease their costs of waste handling (W-Fuel 2011). For this purpose the following practices are implemented:

- Schools and nursing homes have a long tradition for preventing food waste by preparing for people tasty and healthy food in controlled amounts;
- Kinder gardens order their food from food-making services. In order to reduce amount of plate waste, food is distributed in bigger plates (tureens, bowls) and every child can choose sort and amount of meal (e.g. potatoes, meat, soup, milk, salad);
- Large companies in the food industry regulate stocks of quickly spoiling food by offering to their employees meals in companies kitchens, shops, canteens or pay a part of the salary to employees in company's products for reduced price

(W-Fuel 2011).

Large amounts of whey generated in the dairy industry companies are used by cattle breeding enterprises as cheap feed or even free of charge. One of the companies has developed a new beverage, in order to use whey: mixed with fruit juices whey is packaged in 250–1000 ml bottles and marketed through two market chains in Estonia (W-Fuel 2011).

In addition to anaerobic, composting and MBT plants, in order to meet the targets of the Landfill Directive as well as country's own targets, a new incinerator close to Tallinn with a yearly capacity of 220,000 tonnes has been put into operation (Fischer 2013a).

7.2.4 Germany

7.2.4.1 Public Awareness Campaign/Education

In comparison to other discussed countries, Germany is very active in the area of public awareness and educational campaigns aimed at prevention and reduction of food waste. Studies, press reports, radio and TV broadcasts on this subject are released on a regular basis and engaged socio-political attention (Kranert et al. 2012). All these are a reaction of the German society to the independent film by Valentin Thurn, 'Taste the Waste' that shocked audiences about the level of food waste in developed economies (Mccabe and Lieberz 2013).

It is important to note that the problem of food waste is considered on the governmental level, where one of the major players in this area, main initiator and supporter is the BMELV. The Ministry has launched information campaigns against food waste to strengthen the awareness of the value of food (BMELV 2013a).

In addition, the BMELV calls for a national cooperation of the industry, the research community, consumers and associations to combat the problem of food waste. Food producers, traders and large-scale consumers are called upon to avoid waste in a sustainable manner and to use it intelligently. It is recommended to offer at reduced prices food with a best-before date that is about to expire.

To improve cooperation between the industry and food banks and similar projects, the Ministry is currently working on guidelines for donating food to welfare services (BMELV 2013a).

The BMELV supports several projects aimed at providing children with practical tips about the handling of food, for example, by means of the guide to a healthy diet for pupils, which half a million pupils have already acquired in Germany (BMELV 2013a).

7.2.4.2 'Better Appreciation of the Value of Food!' Campaign

In 2010, there was an attempt to raise awareness about food waste with the phrase, 'Better Appreciation of the Value of Food!' It is a common German view that food prices are too low, which leads to wasted food. The BMELV tried to instil more appreciation for food through the campaign showing that food is valuable. However, the campaign has generally failed to catch on (Mccabe and Lieberz 2013).

'Too Good for the Bin' initiative

In 2012, the BMELV has launched 'Too good for the bin' ('Zu gut für die Tonne') initiative (https://www.zugutfuerdietonne.de/). One of the main goals of which is to reach as many people as possible and reduce food waste with joint effort along the entire chain (BMELV 2013c). The campaign mainly targets private households, emphasizing that 65 % of food waste could be avoided. It includes a website, phone App, leaflets at supermarkets, and exhibitions, to inform and stimulate a change of throw-away behaviour. The campaign seeks to inform German households of better planning while grocery shopping, better storage of food, making the most of leftovers, and trusting instinct rather than 'best before' dates on packages. The ultimate goal is to reduce Germany's waste by 50 % by 2020 (Mccabe and Lieberz 2013). The website offers information, useful tips and advices for consumers. It also includes a large database for recipes for leftovers from top chefs and celebrities, and offers an interactive test on how one can reduce food waste in the best way.

According to the survey about 51 % of Germans has heard about the initiative. 26 %stated that they have changed behaviour within the past months and started dealing with food more consciously (BMELV 2013c).

Within the scope of the initiative, various information materials, such as brochures, flyers, postcards, labels and etc., were sent and distributed to student services, associations, schools and food retailers (BMELV 2013d).

7.2.4.3 Nationwide Days of Action Against Food Waste 'We Save Food!'

The BMELV, Slow Food Deutschland e.V. and the Bundesverband Deutsche Tafel organise nationwide days of action against food waste 'We save Food!' ('Wir retten Lebensmittel!'). This includes film screenings, panel discussions, tours and events in schools and kindergartens. Citizens, companies and organizations are invited to participate. Food with flaws is collected from supermarkets and producers and then the leftovers are being processed into a tasty 'best leftover menu'. Highlight of this food-saver-event is a long table in a central square of each city. Citizens are offered delicious leftover meals (BMELV 2013c).

7.2.4.4 Information About the Best-Before Date in the Retail Sector

Together with the Federal Association of the German Retail Grocery Trade, the BMELV launched a nationwide awareness-raising campaign about the 'best-before' date in the retail sector. The aim is to prevent valuable foodstuffs that are still perfectly edible from being thrown away. Over 4 million flyers and information leaflets were distributed in 21,000 supermarkets throughout Germany, providing answers to the most important questions related to the labelling of the 'best-before' and 'use-by' date (BMELV 2013d).

7.2.4.5 Food Donation/Food Banks

Started in 1993, the German Food Banks (Die Tafeln) is a network of more than 900 local/regional units and nearly 1.5 million people (Die Tafeln 2013). It is important to note that the units are legally independent and also organized by individual structures (Hallier 2013). The organisations operate solely through donations (Gentilini 2013). The Banks distribute tens of thousands of tonnes of excess food to people in need, every week (BMELV 2013d). Among the main types of donated food are fruit, vegetables, meat and dairy products (Ujhelyi 2013).

In cooperation with the Austrian Ministry of Agriculture, the BMELV has issued a guide explaining the present legislation to both donors and recipients of food surpluses, in order to facilitate the forwarding of food to social welfare bodies, including the Food Banks (BMELV 2013d).

7.2.4.6 Biological Treatment

Composting plays one of the key roles in treatment of biodegradable waste. According to the Federal German Compost Quality Assurance Organisation (Bundesgütegemeinschaft Kompost e.V.), there are approximately 453 composting plants that together with 120 anaerobic digestion plants, process about 9 million tonnes of bio-waste and produce around 5.7 million tonnes of compost and fermentation products per year (Federal German Compost Quality Assurance Organisation 2014). The country also supports and promotes home composting (Dehoust et al. 2010).

The country is one of the largest markets for MBT plants. The concept of MBT originated in Germany and was established as the waste treatment method (Defra 2013b). There are about 50 MBT plants in operation, in the country (Böhm et al. 2011). These plants pre-treat approximately 6 million tonnes of municipal solid waste annually (Balhar 2010).

Leftover food from restaurants and large kitchens (e.g. canteens, hospitals and refectories), waste from the grocery trade, and production residues from food manufacturers are used by recovery plants (Kern et al. 2012). For example, according to the German statistical office (Statistisches Bundesamt), in 2010, out of 728 thousand tonnes of accumulated biodegradable kitchen (not from households) and canteen waste (food wastage), 99 % was recovered (German Federal statistical Office 2013a, b).

The country is the world market leader in the biogas sector. In 2010, Germany produced around 53 % of the power generated from biogas in Europe. In the same year, only 992 out of 5905 biogas plants generated energy from waste. In 2012, the number of installed biogas plants accounted to almost 7600 (German Biogas Association 2013).

The biogas sector in the country is represented by a variety of companies that also offer services regarding food wastage treatment. Below are examples of such companies:

ReFood, one of the companies of the SARIA Group, a producer of renewable energies and a provider of services to the agricultural and food industries (ReFood 2013a). The company disposes of food and kitchen wastage, used cooking oil and frying fats, as well as out-of-date food (food waste) from catering, trade and industry applications. The annual volume handled is exceeding 450,000 tonnes. Among the ReFood customers are restaurants, canteens and catering companies, bakeries, care home and hospital kitchens, meat markets and butchers' shops, food producers and food retailers. The company uses food and kitchen wastage to generate renewable energy in the form of electricity and heat, and to produce sustainable fertilisers for agricultural use and basic materials for producing environmentally friendly biodiesel (ReFood 2013b).

AAT GmbH, the core competence of the company lies in the systems for treating and processing industrial production waste materials, food wastage, renewable raw materials, agricultural substrates and municipal sludge. The company offers such services as planning, consulting, construction management, equipment supply, start-up support and after-sales service (German Biogas Association 2013).

Schwarting Biosystem GmbH, the company is specialised in the construction and operation of anaerobic digestion plants for processing special organic residue such as food wastage, including expired food products, slop and sewage sludge (German Biogas Association 2013).

7.2.4.7 Additional Activities/Initiatives

Competition between cities and municipalities: Another initiative is a competition among regions within Germany on food waste, organized by the BMELV together with a German firm Projektträger Jülich. The competition between cities and municipalities seeks to identify the region with the best and most innovative practices against food waste (Mccabe and Lieberz 2013).

'Window of the Region' labels: The BMELV has also started the 'Window of the Region' labels that are standardized for all products to clearly state from which region the ingredients originate and where the product has been processed in an effort to cut down on waste in transport. The first products with the regional labels went on sale in various test markets in January 2013 (Mccabe and Lieberz 2013).

Recipes for luscious leftovers: Together with the women's magazine 'BILD der FRAU', the BMELV launched a nationwide competition on the subject of 'luscious leftovers'. The competition looked for the most creative recipes for tasty dishes made from leftover food. Many of the host of competition entries are available for free on the internet (BMELV 2013d).

'Foodsharing' (http://foodsharing.de/): The movement is a response to the V. Thurn's film. Those in the movement seek to collect food that has already been disposed of by supermarkets or share food with one another, refusing to buy food because so much is being wasted. The website connects people, who have a surplus of food, with people, who are searching for food, in this way, food that would have otherwise been thrown away is being given to others (Mccabe and Lieberz 2013).

The founder of 'Foodsharing', Raphael Fellmer, has also set up 'hot spots', where food can be picked up anonymously. He argues that the tonnes of food wasted in Germany could be used to feed people in poor countries (Rettie 2013).

'***Second Bäck***' ('yesterday's pastry') (www.trenntwende.de): A lady buying today's leftover pastry of bakeries and reselling it the following day in two shops in Berlin.

'Culinary Misfits' (http://culinarymisfits.de/): The initiative calls attention to natural appearance of food; question standard specifications by regulation or retail standardization.

EDEKA Supermarket in Bonn processes unsold fruit and vegetables into jam and hot dishes. Food unfit for human consumption is given to pet owners (Braun 2012).

Alliance with the restaurant and catering sector: The German Hotels and Restaurants Association (DEHOGA) is planning to reduce an amount of generated food waste by offering smaller serving sizes in the restaurant and catering sector and by passing food on to charitable organisations such as food banks (BMELV 2013d).

Hospitals looking for new approaches: The largest municipal hospital operator in Germany, Vivantes, managed to reduce the amount of food waste in patient care by as much as 10 % within 12 months, for example by conducting detailed patient surveys and offering flexible product assortments on the wards instead of serving ready-made trays (BMELV 2013d).

Dealing with oversupply: Small bakers and butchers try to recycle not sold products in their daily production into new products or meals. Sometimes bread goes also to horse-feeding or some meat to pets of good customers as a kind of loyalty-program (Hallier 2013).

Online resources: Additional online resources that work with food waste but labelled under 'Wohlfahrt/Charities' are:

- AWO—www.awo.org
- Caritas—www.caritas.de
- Diakonie—www.diakonie.de
- Red Cross—www.drk.de
- Tafeln/Food Banks—www.tafel.de

(Hallier 2013).

7.2.5 Latvia

7.2.5.1 Public Awareness Campaign/Education

According to the study, 'Improving Waste Prevention Policy in the Baltic States. Assessment and Recommendations' (Bremere 2011), for promotion and stimulation of sustainable distribution and consumption, such instruments as awareness raising campaigns, education at schools, technical training courses, green public procurement, marketing, labelling and certification are applied in the country.

7.2.5.2 Food Donation/Food Banks

The Latvian Food bank 'Paēdušai Latvijai'('For a Fed Latvia') was established in 2009, by the charity portal Ziedot.lv in cooperation with regional charity organisations, the Latvian Association of Samaritans, the advertising agency DDB and the international relations agency Nords Porter Novelli (Ziedot.lv 2013). According to the manager of the Food Bank, Agita Kraukle, the organisation helps poor families with children, who cannot get any support from the local government. The food packages contain food products from society and companies donations. Sometimes the Food Bank gets food with close to expired date and with damage packing from food companies. According to the Institute of Development Studies, in 2011, the Latvian Food Bank served 180 thousand people (Gentilini 2013).

In the interview, a representative of one of the major bakery producers in the country stated that the company all overproduction is given to charity.

7.2.5.3 Separate Collection

There is no separate collection of kitchen waste or food waste in Latvia. It is treated together with unsorted municipal waste. Restaurants and hotels are paying for the waste removal according to the normal waste management practice established by waste management agreements (Dzene 2009).

In one of the regions (North Vidzeme), where waste management practices are considered as most developed in the country, the organic fraction of waste is usually partly separated in private households in the rural area. Households in the city area do not separate the organic fraction from the main MSW stream and it goes to landfill sites. From 2009, North Vidzeme Waste Management Organization (ZAAO), the main waste company in the North Vidzeme Region, had started a pilot project of organic waste collection from private food companies (UrbanBiogas 2013).

Experts name the following barriers for establishing waste separation systems:

- low tariffs for the collection of unsorted MSW, which do not motivate the industry and private sector to implement advanced waste management options;
- low income of the inhabitants which does not allow an increase of costs for waste management services;
- low environmental awareness and education of the society;
- lack of legal, financial and administrative instruments for the implementation of advanced waste management options

(UrbanBiogas 2013).

7.2.5.4 Biological Treatment

There are 5 composting large scale plants and 1 anaerobic digestion plant that uses sludge and bio-waste, operating in the country (Moora 2011).

Households in the rural area of the North Vidzeme region compost kitchen waste for private use (UrbanBiogas 2013).

There are companies that offer the services in management of food waste to enterprises. For example, the company 'L&T', the daughter company of the Finnish public listed company 'Lassila and Tikanoja', implements management of three different kinds of waste:

- animal by-products;
- catering industry, food, and vegetable waste;
- oil—plant world origin food oil used for cooking

(Lassila and Tikanoja company 2013).

With regard to regulations for industrial waste management, after the administrative reform in 2010, some local governments have approved local binding regulations. According to these regulations, all companies that produce organic waste must have a contract with a specific organic waste service company. Until now the local governments had not started actual control of these regulations (UrbanBiogas 2013).

7.2.5.5 Additional Activities/Initiatives

Shops and catering companies have contracts for the animal by-products collection with the companies 'Re Cikls' and 'Reneta', but there are no sound data available about volumes and means of utilizations of these waste streams (UrbanBiogas 2013).

Few supermarkets have contracts with some of the by-product processing companies. Thermally processed food wastage according to an agreement with a waste collecting company is delivered to landfill or in very few cases food waste from supermarkets is feed for fur-bearing animals (non-productive animals). In emergency situations (like storm or electricity outage) all food is delivered to a specialised by-product processing company for disposal (Dzene 2009).

Waste from the fruit and vegetable processing companies based on agreements is sold to farms for animal breeding. Waste food oil is used for animal feed production. Wastes from breweries and distilleries are used for animal breeding or as fertilisers. Based on conditions of particular agreements those waste are sold either given for free. Whey is mainly used for cattle and swine breeding. The rest is mixed with manure and applied as a fertiliser. Animal tissue waste and other waste from the meat and fish production are given to certified waste management companies for further processing (Dzene 2009).

7.2.6 Lithuania

7.2.6.1 Public Awareness Campaign/Education

According to a study, for promotion and stimulation of sustainable distribution and consumption, such instrument as awareness and education are applied in the country (Bremere 2011).

7.2.6.2 Food Donation/Food Banks

The Lithuanian Food Bank (www.maistobankas.lt) was started in 2001, as a program run by the charity 'Lithuanian—US Initiatives' supported by Kraft Foods Lithuania. In 2007, the Food Bank program had been transformed into the independent non-profit organization called 'Maisto Bankas'. The Food Bank collects, close to expired date, fresh and perishable food items and distribute them to charities and families in need in 24 Lithuanian towns and cities. In 2010, the organisation collected more than 1,000 tonnes of food that intended to be disposed

(FEBA 2013b). According to the Institute of Development Studies, in 2011, the Lithuanian Food Bank served more than 496 thousand people (Gentilini 2013). In 2012, 52 % of distributed food (1,609 tonnes) came was close to expired date products from retailers and 7 % (217 tonnes) was surplus food from producers (Maisto Bankas 2013).

The Food Bank in collaboration with the 'IKI' chain stores runs a project of minimizing food waste (FEBA 2013b). The retail chain is one of the ten biggest companies in the Baltic States, consisting of over 280 retail centres in Lithuania and Latvia (IKI 2013). Since 2009, the company donates last-expiry-date food products that amounts to about 2,000 tonnes of food per year (Maisto Bankas 2013).

According to the results of the survey, conducted by the Food Bank, 25 % of the interviewed companies stated that they would give food to charities (Ujhelyi 2013).

7.2.6.3 Treatment

Utilisation of food wastage varies, greatly depending on the production facility (Havukainen et al. 2012). In 2005, about 90 % of food wastage (by-products, food remainders) generated in the grain processing companies was dumped. The other 10 % was passed to associations of hunters for wild animal feeding. There are no other alternatives for treatment of this type of waste in the country (Juškaitė-Norbutienė et al. 2013). In the same year, 37 % of animal by-products generated by the fish industry were passed to fur farms and hunters, another 37 % was passed to farmers for animal feeding. About 10 % was used in biogas production. Only about 2 % was burned in the animal by-products treatment factory (Juškaitė-Norbutienė et al. 2013). Annually, about 13 % of the generated amount of animal by-products is passed to various treatment facilities that are neither registered as food preparation waste treatment factories (according to the Lithuanian Environmental Security Agency register) nor as the 3rd category animal by-products treatment factories (according to the Lithuanian State Food and Veterinary Service register) (Juškaitė-Norbutienė et al. 2013).

Despite the available technical capacity for the appropriate treatment of food wastage from meat, fish and dairy industries in the country, there are still cases of ineffective management of this type of waste. The largest part of waste generated by the grains treating industry and bakeries is brought to the landfills (sometimes illegally) for final disposal. The rest is given to farmers for feeding (Juškaitė-Norbutienė et al. 2013).

Liquid waste, generated by producers of dairy products, is not registered and discharged into the general sewage system. The rest is given to farmers, biogas or milk powder production (Juškaitė-Norbutienė et al. 2013).

In 2011, there was a plan to set up 6 anaerobic digestion plants (Moora 2011). With regard to composting, already in 2004, one of the waste management companies has offered services for composting of food wastage (Veidemane et al. 2004).

The food industry and catering companies are faced with a number of problems with regard to the treatment of generated food wastage. Companies are forced to look for the solutions individually that increases costs of waste management. There is also a problem related to low quality of data on waste production, collection, separation and treatment, as well as to classification of sub-types of biodegradable waste. There are cases, when waste producers, collectors and waste treatment companies attribute same waste to completely different types according to the waste list (Juškaitė-Norbutienė et al. 2013).

The authors also noted that it is most difficult to trace the way, types and quantities from the source of generation to the final treatment of waste generated by supermarkets. Principally, supermarkets have contracts with many small waste management companies, but 'waste chains' at these companies often are 'broken' and it is unknown where and how waste is disposed of (Juškaitė-Norbutienė et al. 2013).

7.2.7 Poland

7.2.7.1 Food Donation/Food Banks

The first Polish Food Bank (http://www.bankizywnosci.pl) was founded in 1993, later, in 1997, the Federation of Polish Food Banks was established. Currently, there are 30 Food Banks affiliated with the Federation (FEBA 2013c). The Federation also cooperates with food producers and distributors (Gosiewska 2013). The organization collects rejected food from producers and intermediaries and pass it to the needy (Ciobanu 2013). 16 % of food, distributed by the Food Banks comes from the food production and 2 % is from the food distribution (Polish Food Bank 2012). According to the Institute of Development Studies, in 2011, the Polish Food Bank served more than 3 million people (Gentilini 2013). In 2012, the organisation distributed 50,899 tonnes of food to charities in contact with people in need (FEBA 2013c).

7.2.7.2 Public Awareness Campaign/Education

The Federation is very active in the area of distributing information and promoting educational tools for food waste management. According to the results of the study, conducted by the organization, 75 % of the participants admitted that educational programs are needed to inform about how not to waste food (Ujhelyi 2013; Gosiewska 2013). Already in 2005, the Food Banks have conducted a community wide campaign under the slogan 'Stop the Waste' (Jamiolkowska 2012).

In 2011, the organisation set up the Council for the Rational Use of Food, in order to create conditions that will help in the rational use through the food chain at the level of production, distribution and consumption. The Council works in two main areas: research and education. It also planned to set up a special group to law and regulation. Among the partners of these project are the Warsaw Agricultural University, the Faculty of Human Nutrition and Consumer Sciences, the Warsaw University, the Polish Academy of Sciences, the Institute of Agricultural and Food Economics, the National Food and Nutrition Institute, the Ministry of Agriculture and Rural Development (Gosiewska 2013).

In 2011, the Federation supported a press conference about food waste produced in professional kitchens, organised by the Unilever Food Solutions and Multi Communications companies. Among participants were 18 media representatives of the food service industry, consumers' press and national dailies. During the conference participants discussed the problem of reasonable food waste management, including food waste generated in professional kitchens. In addition, chef of the Unilever Food Solutions company showed how on practice to minimize food waste, generated as a result of overproduction in the food service industry (Multi Communications 2011).

The purpose of the campaign, 'Don't waste food,' is to educate about the environmental impact of food waste and methods of its reduction (Jamiolkowska 2012). In the frame of the campaign an internet platform was developed (www.niemarnuje.pl) which provides practical guide for consumers and how to reduce households food waste (Ujhelyi 2013).

According to the results of the study conducted by the Food Banks, charity organisations expressed their interest in learning about:

- Tax regulations on donating food (e.g. VAT, tax cuts);
- Accounting advise for food waste donations;
- Regulations on food traceability;
- Necessary legal content of contracts with companies.

At the same time, companies, in addition to the mentioned above two first topics, named such aspects as

- Information about charity organisations/food banks and their activities;
- Ways/methods of monitoring charities

(Gosiewska 2013).

7.2.7.3 Separate Collection

The issue of separate collection and sorting is still in the developing phase and represents a major challenge for the country (BiPRO 2011). According to the results of the study 'Economic Analysis of Options for Managing Biodegradable Municipal Waste: Final Report to the European Commission', conducted by Eunomia, in 2007, the proportion of kitchen and yard waste that was treated through separate collection amounted to 0.35 % (Mott 2012). In 2009, there were 173 sorting plants with total capacity of 2,227 thousand tonnes (Deloitte Poland, Fortum, 4P research mix 2011). During the survey, conducted by the Deloitte Poland company, most respondents indicated the availability of containers for sorted waste (76 %) and lack of charge for collection of sorted waste (74 %) as the best motivation for waste sorting, followed by—lower fees for sorted waste (66 %) and the possibility to sell the materials, recovered from sorted waste (63 %) (Deloitte Poland, Fortum, 4P research mix 2011).

7.2.7.4 Treatment

The largest fraction of food wastage from the agri-food industry is generated by slaughterhouses, meat processing companies, dairy companies, refrigeration plants, farms, sugar factories, breweries, distilleries, fruit and vegetable processing companies, and catering facilities.

In 2000, about 98 % of the total amount (12.6 million tonnes) was recovered (Malinska 2004). In the same period, over 84 % of total food wastage from the sugar industry was recovered as a fertilizer and 5 % was deposited at landfills (Malinska 2004). Whey, generated by the dairy industry was used for animal feed or other purposes including production of pharmaceuticals. Only 1 % of the waste from this type of industry was landfilled (Malinska 2004).

In 2008, a large percentage of food wastage generated by industries was utilised, mainly for production of animal feeds and fertilisers (Polish Council of Ministers 2010).

In 2009, the total capacity of thermal waste treatment, mechanical-biological treatment and composting plants amounted to over 1.1 million tonnes per year (Deloitte Poland, Fortum, 4P research mix 2011). In the same year, the list of facilities for treatment of municipal waste included the following (Table 7.11):

Despite the national waste management regulatory regime that strongly favours shifting food wastage to biogas plants, the largest part of household food wastage is still landfilled. One of the reasons is the fact that the Polish legislation does not support such a policy. Household and restaurant food wastage are not covered by the ‘agricultural biogas’ definition and there is no current category of other biogas that includes this material (Mott 2012).

The development of the required infrastructure for waste management faces obstacles related to funding, administration and public omissions. Experts foreseen

Table 7.11 List of municipal waste treatment installations in 2009 (Deloitte Poland, Fortum, 4P research mix 2011)

	Number of installations	Total processing capacity, thousand tonnes
Composting plants for green waste and separately collected organic waste	90	602
Municipal waste incineration plants	1	42
Fermentation plants	3	52
Mechanical biological treatment plants	11	412

that new technologies for energy recovery from waste will be comprehensively introduced. However, prognoses suggest that only some of the planned incinerators will be built. Investments are strongly focusing on MBT and Refuse-derived fuel (RDF) technologies (BiPRO 2011). It is planned to set up between 87 and 97 composting and fermentation plants, 28-30 MBT plants and 27 sorting plants for municipal waste treatment (Deloitte Poland, Fortum, 4P research mix 2011).

According to the programme of development of biogas plants in Poland by 2020, presented by the Polish Ministry of Economy in 2011, it is planned to use by-products from agriculture and waste from the agricultural and food industry for biogas production (Radziszewski 2011).

The results of the survey, conducted by the Deloitte Poland Company showed that 71 % of the respondents gave the highest approval to 'modern waste incineration plants' as to a method of treatment and disposal of waste. This method was followed by composting and biodegradation that were approved by 67 % and 60 % of the interviewees, respectively (Deloitte Poland, Fortum, 4P research mix 2011).

7.2.7.5 Additional Activities/Initiatives

The Unilever Food Solutions company, in their 'quest' to find solutions to the problem of food waste overproduction, has engaged in cooperation with food service industry representatives and other organizations, with the aim of alleviating the consumers' concerns and helping the restaurants and food retailers to make their kitchens more sustainable. The company provides kitchen operators and chefs with professional guidance on how to be more effective in managing food supplies and reducing food waste (Multi Communications 2011).

All waste, generated by the Wrigley factory, is recycled or otherwise diverted from landfill. The Mars Poland company has already achieved the goal of zero waste to landfill (Mars 2012). Also two Wrigley facilities have been designed to capture methane from their waste treatment operations and redirect it to fuel boilers that heat water (Food Drink Europe 2012).

Today, in Poland, only producers who donate food are spared the VAT tax, while retailers are not. Therefore, the Federation argues for a scrapping of the VAT tax for food donations from any company (Ciobanu 2013).

7.2.8 Sweden

7.2.8.1 Cooperation

The topics of food waste and food wastage have been on the agenda in Sweden for some years. One of the outcomes is establishment of the network 'SaMMa' (Samverkegruppen för minskat matavfall/a group for food waste reduction). It is a platform for cooperation of 30 organisations that represent various sectors, for example, the national authorities are represented by the Ministry of Environment, the Swedish Environment Protection Agency (Naturvårdverket), the National Food Agency (Livsmedelsverket), the Ministry of Agriculture and the Ministry of Rural affairs, whereas, the hospitality sector by the Swedish Rural Economy and Agricultural Societies (Hushållningsselskapet) and the Swedish Hotel and Restaurant Association (SHR). The group discusses issues on food waste and exchanges experiences and knowledge. This cooperation is considered as the 'driver' for reducing food waste in the country (Marthinsen et al. 2012).

7.2.8.2 Public Awareness Campaign/Education

In 2009, a multistep campaign 'Food weighing in schools', in the Karlskrona municipality was launched. During 2–3 weeks, the food, thrown away by the municipal primary and secondary schools, was weighed. Based on the results, teachers and meal personnel, aided by posters, brochures and other information media, put forward such topics as 'eat well—feel well', 'eat more vegetables' and 'do not throw away food' (BIO Intelligence Service, Umweltbundesamt and Arcadis 2011).

Between November 2009 and March 2010, the Swedish Eurest company, one of the members of the Compass group, food and support services company (Compass group Sweden 2014), has initiated a campaign, aimed at reduction of food waste from the kitchen and guests in 120 canteens and cafés in 44 municipalities. For this purpose, 25 restaurants and 2 coffee shops in 15 different places across the country weighed and measured waste from their preparations and from the guests during a day. The results were published on posters for guests and staff in the restaurants. In addition, the restaurants provided information regarding the negative impacts of food waste and advices on what can be done about it (BIO Intelligence Service, Umweltbundesamt and Arcadis 2011). For the project, Eurest produced a 10-measure list to reduce food waste and related waste, for both, guests and staff and with this to improve production and planning of the menu (Karlskrona Municipality 2010). At the end of the project, Eurest reported a reduction in food waste per serving from 130 to 101 g i.e. by 23 % (Marthinsen et al. 2012).

Another informational national campaign, 'Basta for food waste,' ('Basta till matsvinnet'), regarding the problem of food waste was initiated by the Swedish politician, a member of the European Parliament, Anna-Maria Corazza Bildt (Westregård 2012).

In 2010, the Swedish National Food Agency (Livsmedelsverket) published reports to increase knowledge on the issue of food waste. The organization is also working on integration the food waste issue into the general work towards food services at schools and social care (Marthinsen et al. 2012). On their website, the Agency offers advice/recommendations on how to store food properly, use leftovers efficiently and learn about the real meaning of 'best-before' dates and also provides information about negative environmental consequences of food waste (Swedish National Food Agency 2013).

In 2012, Djupfrysningsbyrån, which acts as a forum for companies that work with refrigerated and deep-frozen food in cooperation with various food companies, released a book that contains guidelines on how to reduce food waste (Westregård 2012).

Based on the results of the conducted studies that also included picking analyses and attitude surveys, the Swedish Organisation for Local Enterprises (KfS) released a consumer guide, 'Do not throw away food'('Släng inte maten') that is available for downloading from their campaign website (www.slangintematen.se) (Westregård 2012).

The Lantmännen group that owned by 33,500 Swedish farmers and operates in 22 countries (Lantmännen Group 2014), regularly promotes waste reduction activities by commercial messages for consumers under the slogan 'From farm to fork' ('Från jord till bord'). They have also released the book, 'Be careful with food,' ('Var rädd om maten') about alternative ways of treating food instead of throwing it (Westregård 2012).

The Swedish Rural Economy and Agricultural Societies (Hushållningsselskapet), with support of the Swedish Board of Agriculture, conduct training activities for the municipal operated kitchens as part of the campaign for a reduction of food waste. The training is based on workshops with few participants. In 2011, 200 people have been trained at 12 locations across the country (Marthinsen et al. 2012).

The Sysav company, runs 'Eco-cycle Plan', one of the goals of which, is to inform and spread knowledge about how to reduce the volume of waste. Beginning with a seminar in 2011, a major initiative 'Reducing food waste' followed by training courses for people who work with food in schools, elderly care and other municipal catering operations, focusing on facts about food waste, ractical exercises and tips on how to reduce it in the kitchen. Furthermore, a network has been formed involving people with responsibility for food and catering, where they can share experiences and pass on knowledge (Sysav Biotec 2012).

According to the SHR, increased knowledge and competence, both, among players in the market and among food safety inspectors might help to stimulate more actions and thus, to achieve the future targets set by the government (Marthinsen et al. 2012).

7.2.8.3 Food Donation/Food Banks

The Allwin company, established in 2010, in order to 'help companies taking care of their overproduction' (Westregård 2012), redistributes excess from food production and other products on behalf of private and public organizations, helping people with their needs (Marthinsen et al. 2012).

The ICA Sweden group has a central agreement with the Salvation Army and several local partnerships with charities to donate food from their warehouses that is soon expiring or cannot be sold in stores because of damaged packaging. The donated products meet the same basic food safety requirements as ICA's other products. Food that has passed its 'best-before' date is never donated (ICA Group 2013).

In regard with the establishment of food banks, there have been obstacles, linked to delivering the right type of food to the right place, within a certain time. Currently, 'food responsibility' has been clearer described by the EPA and the National Food Agency that might help in overcoming existing barriers. To avoid problems arising from the cooperation with food banks, stores prefer to establish a kitchen on their own, for example, as it was organised by ICA Malmborgs in Lund (Stenmarck et al. 2011).

7.2.8.4 Separate Collection

In Sweden, the food wastage management system differs among municipalities. Food wastage falls under the category 'waste similar to household waste' and is looked upon as the responsibility of a municipality, as well as waste generated by stores (Stenmarck et al. 2011).

Separation and processing of food wastage have been taking place at various locations in the country for several years (Sysav Biotec 2013). According to the Swedish waste management report 2012, collection of source-separated food wastage is on the increase. Currently, about 60 % of municipalities have introduced collection systems for source-separated food wastage. About 20 of them only collect food wastage from restaurants and large-scale kitchens, while the remaining municipalities have systems for households as well. An additional 70 municipalities are planning to follow suit (Avfall Sverige 2012).

There are two types of collection systems of household waste. It can be collected either as a mixed fraction intended for waste-to-energy recovery or in separate fractions—one for food wastage and one for combustible waste, by a multi-compartment bin or by optical sorting.

The multi-compartment system is a system in which different fractions are separated into separate containers. Optical separation is collection of waste into different coloured bags that are put into the same container. The system is now being used in more municipalities than before (Avfall Sverige 2012).

Large-scale kitchens and restaurants can also use disposal grinders, whereby food wastage is collected in a tank (Sysav Biotec 2013).

Many municipalities introduced the voluntary collection of food wastage by using a fee as a means of control. In other words, those who choose a food wastage subscription pay a lower fee than those who choose to deposit mixed waste (Avfall Sverige 2012).

On other hand, bin-based collection of food wastage is often subsidised, and therefore less expensive for the waste generator/property manager, that makes it to consider as the preferred system if no extra prerequisites exist (Millers-Dalsjö and Lundborg 2012).

7.2.8.5 Pre-Treatment of Food Wastage

Source- separate collection is not the only step required before food wastage is sent to waste treatment plant. There are additional pre-treatment steps of food wastage, implemented in the country. According to the Sysav Biotec company, effective pre-treatment of food wastage is important, to ensure that subsequent treatment in a biogas plant works well (Sysav Biotec 2013).

For example, the grinder-to-tank system is used to collect food wastage generated from restaurants and catering kitchens. A grinder at a restaurant not only reduces the volume of waste, but also the amount of reject from the biogas digester since it directly signals to the staff if poorly separated waste is put into the grinder (Millers-Dalsjö and Lundborg 2012).

Pre-treated food waste is technically more suitable for any existing digester, since it is pumped directly to the digester for production of biogas. However, such system has relatively high investment costs that slow down its increased further implementation (Millers-Dalsjö and Lundborg 2012).

In 2009, the Sysav South Scania Waste Company has built a facility to process food wastage into biogas and biofertilisers. The production process consists of two steps. The first step is pre-treatment of waste. Sysav's plant can receive three types of food wastage: pumpable liquid food wastage (e.g. waste from disposal grinders and fat separation sludge), pre-packed liquid food waste (e.g. milk and juice in cartons), and separated solid food wastage (standard, separated food wastage from households, restaurants and large-scale kitchens such as school canteens). Pre-treated waste leaves the plant in the form of pumpable slurry.

The second step is production of biogas and biofertilisers. The slurry is transported to a biogas plant, where, both biofertilisers and a renewable transportation fuel, biogas, are produced through digestion (Eco-Innovation Observatory 2013). The residual product in the form of combustible waste, generated during the pre-treatment step, apart from the slurry, is used for energy recovery in Sysav's waste-to-energy plant (Sysav Biotec 2013).

During the years 2010–2012, the company pre-treated between 15,500 and 28,600 tonnes of food wastage from the region's households, restaurants and food processing companies, in the pre-treatment plant (Table 7.12) (Sysav Biotec 2012).

7.2.8.6 Treatment

The country is very ambitious in its plans regarding the solutions of the problem of food wastage. One of the environmental targets regarding waste management on the national level is to reduce the amount of food wastage by 20 % between years 2010

Table 7.12 Treated wastage at Sysav's pre-treatment plant, in 2010–2012, in tonnes (Sysav Biotec 2012)

Year	2010	2011	2012
Treated food wastage (tonnes)	15,500	24,200	28,600

and 2015 (Stare et al. 2013). It was calculated that this 20 % target is equal to a 35 % reduction in food waste. The target comprises all sectors (Marthinsen et al. 2012).

In addition, the new Swedish waste management plan for the years 2012–2017 ('Från avfallshantering till resurshushållning—Sveriges avfallsplan 2012–2017') states that by the year 2018, 50 % of food wastage from households, large-scale kitchens, stores and restaurants being separated and treated biologically to recover plant nutrients and at least 40 % being treated to recover energy (Avfall Sverige 2012).

In 2008, approximately 20 % of food wastage was treated biologically in various compost and biogas plants (Green Advisor 2010).

In 2009, about 21 % of food wastage from households, restaurant and shops was recovered by biological treatment. The corresponding figure for one of the counties (the county of Västmanland) was 60 % (Guziana et al. 2012).

Between the years 2010 and 2012, there was a slight increase in the amount of household food wastage treated biologically, particularly through anaerobic digestion, which also influenced on the decrease of the amount of wastage being composted (Figs. 7.50 and 7.51). According to the Bernstad and la Cour Jansen 2011, the most prevailing treatment method for source-separated household food wastage is centralised, large scale plants, where food wastage is co-digested with other types of organic waste (Bernstad and la Cour Jansen 2011).

Year 2012—Waste from the food industry, slaughterhouses, etc. is not included.

Year 2011: Food wastage that undergoes anaerobic digestion at purification plants—Source: (Avfall Sverige 2013)

However, in the wholesale and retail sectors, large amounts of food wastage still go to incineration mixed with other wastes (Stenmarck et al. 2011). The results of

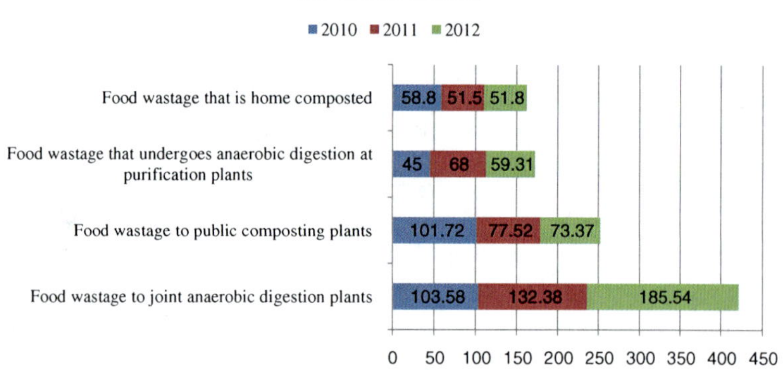

Fig. 7.50 Biological treatment of household food wastage (thousand tonnes) (Avfall Sverige 2012). *Year 2012* Waste from the food industry, slaughterhouses, etc. is not included. *Year 2011* Food wastage that undergoes anaerobic digestion at purification plants. *Source* (Avfall Sverige 2013)

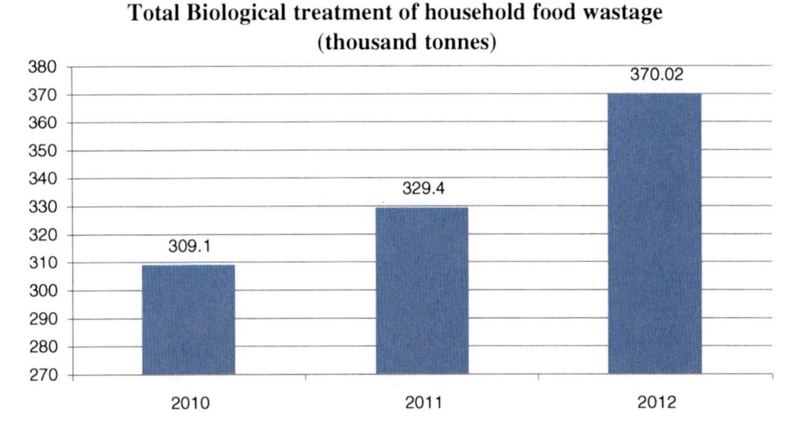

Fig. 7.51 Total Biological treatment of household food wastage (thousand tonnes) (Avfall Sverige 2012)

the study in six Swedish retail stores showed that 33 % of generated retail food waste is incinerated and the rest is composted (Eriksson et al. 2012). For example, food wastage that is not treated biologically, used as fuel in Sysav's incinerator, primarily to produce district heating (Corvellec 2012).

7.2.8.7 Additional Activities/Initiatives

Retail Sector

The retail sector is very active in conducting initiatives aimed at reduction of food waste. Among on-going initiatives are the following:

- Selling wonky fruits and vegetables to a lower price with the arguing that it will taste the same;
- Trying to influence the suppliers (or their own suppliers) to provide smaller amount of certain products;
- Participating in research project regarding food waste reduction;
- Trying to influence certain producers to achieve a more suitable packaging for the product (can be both in terms of size but also in terms of optimised packaging in order to get more out of it or to make it last through the transport);
- Keeping good track of the food in warehouses and set regulations for when (how close before the best before date) products should be marked as 'distribute soon'—in order to maximise the shelf life in a store

(Stenmarck et al. 2011).

The ICA Sweden Group is one of three dominators of the Swedish wholesale and retail food market, with the share of 49.4 % (Dahlbacka 2012). In 2012, the total waste volume generated by the company (72,705 tonnes) was treated in the following way (Fig. 7.52).

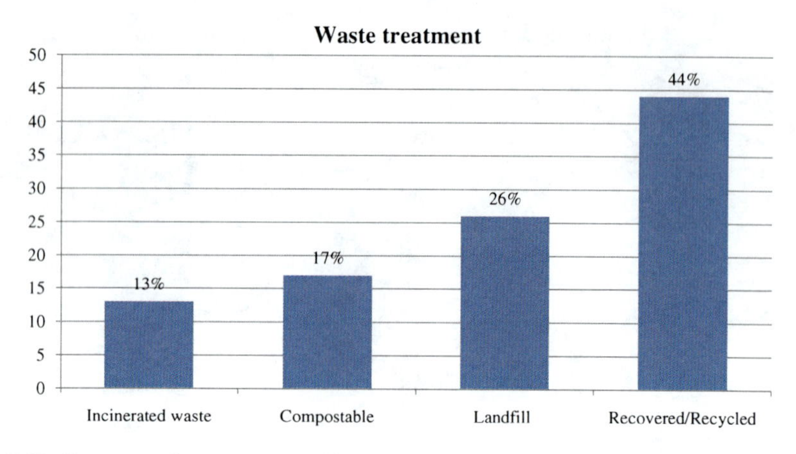

Fig. 7.52 Treatment of waste generated by the ICA Group in 2012 (ICA Group 2013)

The Group has introduced a concept ‘Eat Soon’ which allows customers to buy products that otherwise would go to waste, at sharply discounted prices (ICA Group 2013). The company also developed a labelling system. A green label with the text ‘to be used quickly’ is stuck to products nearing their ‘use-by’ date, and the price of such products is also reduced (RREUSE 2010).

Axfood is food retail and wholesale business that conducted through the wholly owned store chains Willys, Hemköp and PrisXtra and through Dagab and Axfood Närlivs, respectively. Generated food wastage is sorted for biogas production in the municipalities that have suitable biogas plants. One of the company’s ambition is to be able to use food wastage from all stores for biogas production (Axfood Group 2012).

According to the interviews, the following initiatives have already been taken within the sector:

- Optimisation of the selling of products. Products with short shelf-lives, price reduced products with short shelf-life are put on display;
- Management of orders in relation to sale—better prediction of the needs of customers;
- Better handling of food—for example, keeping products stored and exposed under right temperature and light, optimal packaging-size, etc.;
- Education of personnel—how and when to place orders, how to handle and store food, and knowledge of the best practices and the routines for treatment of food waste

(Stenmarck et al. 2011).

Hospitality and Service Sectors

Initiatives taken by the hospitality sector are:

- Participation in the survey initiated by the 'Konsumentföreningen' (consumer association) on food waste and also the attitudes related to the use of doggy bags within restaurants in Sweden;
- Participation in the European Week for Waste Reduction (EWWR) and sharing of experiences within the hospitality sector;
- Projects on food waste reduction, 'Mindre Matsvinn', with a web page, newsletters and training of staff within municipal operated kitchens;
- Several surveys and projects within schools

(Marthinsen et al. 2012).

McDonalds in Sweden operates 221 restaurants and serves 440,000 guests daily. The company is the leader within the hospitality sector. McDonalds sort their food waste for years into two fractions:

- Food waste before food is prepared;
- Food waste from prepared, not sold food.

The company also involves their guests into the process. At 60 restaurants, customers are offered to sort their food waste by themselves (Marthinsen et al. 2012).

Government

The National Food Agency works on improvements in the legislation in order to reduce food waste (Marthinsen et al. 2012).

Digestate that is based on source-separated food wastage, is certified in accordance with the Swedish Waste Management and Recycling association (Avfall Sverige) certification system that includes the Swedish organic label 'KRAV' and the Swedish Seal of Quality (Svenskt Sigill Kvalitetsråd) (Avfall Sverige 2012).

The Swedish Waste Management and Recycling Association (Avfall Sverige) published a report 'Help for introduction of system for collecting sorted food wastage'. The report describes possible collection systems with technical descriptions and examples of appropriate collection vehicles, common collection intervals, increased cost for containers, information, and more. It also contains a guide which specifies what a municipality should consider when establishing a scheme for collection of food wastage to be introduced. The guide is also published separately, under the name 'Manual for implementing collection of sorted food wastage' (Guziana et al. 2012).

With support from the local investment programme, the Municipality of Linköping, in 2001, has started a project aimed at collection and digestion of waste from canteens and restaurants.

In the frame of the project, the municipality installed a waste macerator at the university hospital and in two school canteens and restaurants. A tank to store the macerated waste was also installed. Waste was then transported to the Linköping biogas facility.

The municipality also installed a central screw press facility in its Technical Department. The facility handles liquid foods that have passed their ‘best-before’ date from manufacturers and wholesalers.

The Technical Department plans to purchase a central waste macerator to enable it to handle large volumes of solid waste from wholesalers, such as fruit and vegetables.

The implementation of the project has led to a decrease of the quantity of waste incinerated, an increase of biogas production and of the quantity of biofertiliser going to agriculture (Swedish Environmental Protection Agency 2009).

Thus, the stages of the food supply chain where the biggest amounts of generated food are different in each country. The same is applied to the composition of food waste; it varies from by-products, leftovers to unopened food items. The applied treatment methods belong to both, the upper and lower parts of the food waste hierarchy. The prevalence of one or another method in a country, mostly, depends on available resources, key actors and public activities.

References

Avfall Sverige (2012). *Swedish waste management 2012*. Retrieved from http://www.avfallsverige.se/fileadmin/uploads/Rapporter/SWM2012.pdf

Avfall Sverige (2013). *Svensk Avfallshantering (Swedish waste management) 2013*. Retrieved from http://www.avfallsverige.se/fileadmin/uploads/Rapporter/svensk_avfallshantering_2013.pdf

Axfood Group (2012). *Axfood sustainability report 2012*. Retrieved from http://www.axfood.se/Global/Milj%C3%B6 och ansvar/Axfood_HR_2012.pdf

Balhar, M. (2010). Status quo and perspectives of mechanical-biological waste treatment (MBT) in Germany. *Euro Eco 2010* (pp. 1–12). Germany: Hannover.

Barilla Center for Food and Nutrition (2012). *Food waste: causes, impacts and proposals*. Retrieved from http://www.barillacfn.com/wp-content/uploads/2012/11/WEB_ENG.pdf

Bernstad, A., & la Cour Jansen, J. (2011). A life cycle approach to the management of household food waste - A Swedish full-scale case study. *Waste management*, *31*(8), 1879–1896. Retrieved November 20, 2013, from http://www.ncbi.nlm.nih.gov/pubmed/21511455

BIO Intelligence Service (2010). *Preparatory study on food waste across EU 27*. Retrieved from http://ec.europa.eu/environment/eussd/pdf/bio_foodwaste_report.pdf.

BIO Intelligence Service, Umweltbundesamt and Arcadis (2011). *Guidelines on the preparation of food waste prevention programmes*. Retrieved from http://ec.europa.eu/environment/waste/prevention/pdf/prevention_guidelines.pdf

BiPRO (2011). *Roadmap for Poland (PL)*. Retrieved from http://ec.europa.eu/environment/waste/framework/pdf/PL_Roadmap_FINAL.pdf

BiPRO (2012a). *Country factsheet Estonia (EE)*. Retrieved from http://ec.europa.eu/environment/waste/framework/pdf/EE factsheet_FINAL.pdf

BMELV (2013a). BMELV information campaign Too good for the bin—Strategies against food waste. Retrieved August 19, 2013, from http://www.bmelv.de/SharedDocs/Standardartikel/EN/Food/TooGoodForTheBin.html

BMELV (2013b). Food should not be thrown away. Retrieved August 19, 2013, from http://www.bmelv.de/SharedDocs/Standardartikel/EN/Food/FoodShouldNotBeThrownAway.html

BMELV (2013c). *Too good for the bin—an initiative of the Federal Government to avoid food waste*. Retrieved from https://www.zugutfuerdietonne.de/uploads/media/TooGoodForTheBin.pdf

BMELV (2013d). Too good for the bin—First positive results of our alliance against food waste. Retrieved August 19, 2013, from http://www.bmelv.de/SharedDocs/Standardartikel/EN/Food/ZgfdT_ZwischenbilanzMai2013.html

Boerefijn, P. (2012). How much food goes to waste, pp. 1–21. Retrieved from http://foodweb.ut.ee/s2/109_185_70_How_much_food_goes_to_waste.pdf.

Böhm, K., Tintner, J., & Smidt, E. (2011). Modelled on nature—biological processes in waste management. *Integrated Waste Management, 1*(153–178). Retrieved from http://www.intechopen.com/books/integrated-waste-management-volumei/modelled-on-nature-biological-processes-in-waste-management

Braun, S. (2012). Food Waste. Report on the situation and recent activities in Germany, (October). Retrieved from http://ec.europa.eu/dgs/health_consumer/dgs_consultations/docs/ag/summary_ahac_05102012_3_susanne_braun_en.pdf

Bremere, I. (2011). *Improving Waste Prevention Policy in the Baltic States*. Retrieved from http://www.bef-de.org/Members/befadmin/publikationen/WP2011/activity4-1-1_recommendations_waste-prev.pdf.

Ciobanu, C. (2013). The Secret Treasure of Food Waste. *Inter Press Service*. Retrieved August 5, 2013, from http://www.ipsnews.net/2013/06/the-secret-treasure-of-food-waste-2/

Cofresco (2013). Every fifth bag of groceries goes to waste. *Save Food—an initiative by TOPPITS®*. Retrieved December 12, 2013, from http://www.cofresco.de/en/unternehmen/save-food.html

Compass group Sweden (2014). About us. Retrieved January 27, 2014, from http://www.compass-group.se/In-English/

Corvellec, H. (2012). *Normalising Excess: An Ambivalent Take on the Recycling of Food Waste into Biogas*, Helsingborg. Retrieved from http://www.ism.lu.se/fileadmin/files/rs/wp/WP_15_NOV_2012.pdf

Dahlbacka, B. (2012). *Retail Food Sector Report for Sweden and Finland*. Retrieved from http://gain.fas.usda.gov/Recent GAIN Publications/Retail Food Sector Report for Sweden and Finland_Stockholm_Finland_9-6-2012.pdf

Defra (2013b). *Mechanical Biological Treatment of Municipal Solid Waste*. Retrieved from https://www.gov.uk/government/uploads/system/uploads/attachment_data/file/221039/pb13890-treatment-solid-waste.pdf

Dehoust, G., et al. (2010). *Development of scientific and technical foundations for a national waste prevention programme*. Retrieved from http://www.umweltdaten.de/publikationen/fpdf-l/4044.pdf

Deloitte Poland; Fortum; 4P research mix (2011). *Waste management in Poland Challenges in view of EU requirements and legislative changes; public opinion and prospects*. Retrieved from http://www.deloitte.com/assets/Dcom-Poland/Local Assets/Documents/Raporty, badania, rankingi/pl_Report_Waste management in Poland.pdf

Die Tafeln (2013). Zahlen and Fakten (Facts and Figures). Retrieved August 14, 2013, from http://www.tafel.de/die-tafeln/zahlen-fakten.html

Dzene, I. (2009). *Biogas Potential in Latvia. Summary Report*, Riga, Latvia. Retrieved from http://www.big-east.eu/downloads/IR-reports/ANNEX 2-13_WP2_D2.2_Summary-Latvia_EN.pdf

Eco-Innovation Observatory (2013). Pre-treatment plant for food waste. Retrieved November 17, 2013, from http://www.eco-innovation.eu/index.php?option=com_content&view=article&id=627:sysav&catid=75:sweden

FAO Economic and Social Development Department (2001). *Food balance sheets—A Handbook*, Rome. Retrieved from http://fao.org/docrep/fao/011/x9892e/x9892e00.pdf.

Eisner, S. (2012). Actions to tackle food waste. Retrieved from http://ostfoldforskning.no/uploads/dokumenter/Food Waste juni 2012/Actions to tackle Foodwaste-Allwin.pdf

Engström, R., & Carlsson-Kanyama, A. (2004). Food losses in food service institutions Examples from Sweden. *Food Policy, 29*(3), 203–213. Retrieved December 13, 2013, http://linkinghub.elsevier.com/retrieve/pii/S030691920400020X

Eriksson, M., Strid, I., & Hansson, P.-A. (2012). Food losses in six Swedish retail stores: Wastage of fruit and vegetables in relation to quantities delivered. *Resources, Conservation and Recycling, 68*, 14–20. Retrieved from http://dx.doi.org/10.1016/j.resconrec.2012.08.001

Estonian Environment Information Centre (2010). *Estonian environmental review 2009*, Tallinn. Retrieved from http://www.keskkonnainfo.ee/publications/4263_PDF.pdf

Estonian Ministry of the Environment (2007). *Estonian Environmental Strategy 2030*. Retrieved from http://www.envir.ee/orb.aw/class=file/action=preview/id=1101230/inglisekeelne.pdf

Eurobarometer (2011). *Attitudes of Europeans towards resource efficiency*. Retrieved from http://ec.europa.eu/public_opinion/flash/fl_316_en.pdf.

Europe, Food Drink. (2012). *Environmental sustainability vision towards 2030*. Brussels: Achievements, Challenges and Opportunities.

FAO (2013). *FAO Statistical Yearbook 2013. World Food and Agriculture*, Rome. Retrieved from http://www.fao.org/docrep/018/i3107e/i3107e00.htm.

FEBA (2013a). Lithuania. *Countries*. Retrieved January 17, 2014, from http://www.eurofoodbank.eu/portail/index.php?option=com_content&view=article&id=51:lituanie&catid=13&Itemid=26&lang=en

FEBA (2013b). Poland. *Countries*. Retrieved October 7, 2013, from http://www.eurofoodbank.eu/portail/index.php?option=com_content&view=article&id=53:pologne&catid=13&Itemid=26&lang=en

Federal German Compost Quality Assurance Organisation (2014). Bioabfall (Bio-waste). *Zahlen/Daten/Fakten*. Retrieved January 27, 2014, from http://www.kompost.de/index.php?id=271

Fischer, C. (2013a). *Municipal waste management in Estonia*, Retrieved from http://www.eea.europa.eu/publications/managing-municipal-solid-waste/estonia-municipal-waste-management

Foresight (2010). *How can waste reduction help to healthily and sustainably feed a future global population of nine billion people?* London. Retrieved from http://www.bis.gov.uk/assets/foresight/docs/food-and-farming/workshops/11-608-w4-expert-forum-reduction-of-food-waste.pdf

Foresight (2011). *The Future of Food and Farming: Challenges and choices for global sustainability*, London. Retrieved from http://www.bis.gov.uk/assets/foresight/docs/food-and-farming/11-546-future-of-food-and-farming-report.pdf

Gentilini, U. (2013). *Banking on Food: The State of Food Banks in High-income Countries*. Retrieved from http://opendocs.ids.ac.uk/opendocs/bitstream/handle/123456789/2323/Wp415.pdf?sequence=1

German Biogas Association (2013). *Biogas an all-rounder. New opportunities for farming, industry and the environment*. Retrieved from http://www.german-biogas-industry.com/index.php?id=3

German Federal statistical Office (2013). *Environment. Waste Balance 2011*, Wiesbaden. Retrieved from https://www.destatis.de/EN/FactsFigures/NationalEconomyEnvironment/Environment/EnvironmentalSurveys/WasteManagement/Tables/WasteBalance2011.pdf?__blob=publicationFile

German Federal Statistical Office (2013). Foreign trade. Retrieved December 22, 2013, from https://www.destatis.de/EN/FactsFigures/NationalEconomyEnvironment/ForeignTrade/TradingGoods/Tables/ImportsExports.html

Gosiewska, M. (2013). *FoRWaRD Regional Report*, Retrieved from http://foodrecoveryproject.eu/wp-content/uploads/2012/11/Regional-report-Poland.pdf

Green Advisor (2010). *Food waste that fuels cars*. Retrieved from http://www.afconsult.com/upload/Marketing_Web_Sites/GAR/GAR References/GAR201002_FoodFuel.pdf

ICA Group (2013). *The ICA Group's Annual Report and Sustainability Report 2012*. Retrieved from http://reports.ica.se/ar2012en/Materiale/Files/ICA+annual+report+and+sustainability+report+2012_opt.pdf

Gustavssona, J., & Stagec, J. (2011). Retail waste of horticultural products in Sweden. *Resources, Conservation and Recycling, 55*(5), 554–556. Retrieved from http://dx.doi.org/10.1016/j.resconrec.2011.01.007

Guziana, B., et al. (2012). *Waste-to-energy in a Polish perspective*. Retrieved from http://www.diva-portal.org/smash/get/diva2:574913/FULLTEXT01.pdf

Hallier, B. (2013). *FoRWaRD Regional Report Germany*. Retrieved from http://foodrecoveryproject.eu/wp-content/uploads/2012/11/Regional_report-Germany.pdf.

Havukainen, J., et al. (2012). Potential of energy and nutrient recovery from biodegradable waste by co-treatment in Lithuania. *Waste management and research*, *30*(2), 181–189. Retrieved November 24, 2013, from http://www.ncbi.nlm.nih.gov/pubmed/22105761

Horkova, D. (2013). In Belarus, inspectors are sent to stores to look for spoiled food products (V Belarusi inspektorov zasylajut v magaziny iskat' isporchennye produkty). *Economy*. Retrieved January 4, 2014, from http://mir24.tv/news/economy/8338628

IKI (2013). IKI. Retrieved January 17, 2014, from http://www.iki.lt/en.php

Information Analysis Portal of the Union State (2013a). On the implementation of the Union State program Improving the efficiency of food production by processing its waste with use of advanced technologies and equipment (O realizacii programmy sojuznogo gosudarstva « Povyshenie jeffektivnosti pishhevyh proi. Retrieved January 2, 2014, from http://old.soyuz.by/ru/?guid=139947

Information Analysis Portal of the Union State (2013b). The results of the Union State program "Improving the efficiency of food production by processing its waste with use of advanced technologies and equipment" (Rezul'taty programmy sojuznogo gosudarstva Povyshenie jeffektivnosti pishhevyh proizvodstv za sc. Retrieved September 22, 2013, from http://old.soyuz.by/ru/?guid=140065

Jamiolkowska, K. (2012). Stop Wasting Food—Edelman Poland Helping to Feed the Poor. Retrieved August 5, 2013, from http://www.edelman.com/post/stop-wasting-food-edelman-poland-helping-to-feed-the-poor/

Joint European Union and United Nations Development Programme Project (2013) Waste management. Retrieved from http://www.greenlogic.by/accomplishment_p_1.html

Juškaitė-Norbutienė, R., Miliūtėand, J., & Česnaitis, R. (2013). Bio-Degradable Waste and By-Products from Food Industry Management Systems in Lithuania: Analysis, Problems and Improvement Possibilities. *Environmental Research, Engineering and Management*, *42*, 60–69. Retrieved from http://www.apini.lt/files/4f1db7384c3f18942a4a516f0d5f10ca

Kalinovskaya, T. (2013). Energy available to all (Jenergija, dostupnaja vsem). *Belorussians and Market (Belorusy i Rynok)*. Retrieved January 2, 2014, from http://www.belmarket.by/ru/251/65/19991/Энергия-доступная-всем.htm

Karlskrona municipality (2010). *Eurest Services AB, Pre-waste project*. Retrieved from http://www.prewaste.eu/index.php?option=com_k2&view=item&id=410:106_karlskrona_eurest_food_waste_draft&Itemid=94

Kern, M., et al. (2012). *Ecologically sustainable recovery of bio-waste*, Berlin. Retrieved from http://www.bmu.de/fileadmin/Daten_BMU/Pools/Broschueren/bioabfaelle_2012_en_bf.pdf

Kranert, M. et al. (2012). *Determination of discarded food and proposals for a minimization of food wastage in Germany. Abridged Version*, Stuttgart. Retrieved from http://www.bmelv.de/SharedDocs/Downloads/EN/Food/Studie_Lebensmittelabfaelle_Kurzfassung.pdf;jsessionid=D16102240D2BF5C765C01411351A57ED.2_cid367?__blob=publicationFile.

Kranert, M., Barabosz, J., & Hafner, G. (2013). Feature. Food Waste–how much and how to prevent? - The situation in Germany. IGCS Bulletin, 2(4), pp.5–9. Retrieved from http://www.igcs-chennai.org/wp-content/uploads/2013/10/IGCS-Bulletin-October-20131.pdf

Lantmännen Group (2014). About Lantmännen. Retrieved January 27, 2014, http://lantmannen.se/en/Start/About-Lantmannen/

Lassila and Tikanoja company (2013). Management of Food Waste. *Services and Products*. Retrieved December 9, 2013, from http://www2.l-t.lv/index.php?&454

Lysuho, N. A., & Eroshina, D. M. (2011). *Production and Consumption waste and its impact on environment. Monograph (Othody proizvodstva i potreblenija, ih vlijanie na prirodnuju sredu. Monografija)*. Minsk: International Sakharov Environmental University. Retrieved from http://www.iseu.by/m/12_0_1_64293.pdf.

Maisto Bankas (2013). Lithuanian Food Bank. Retrieved from http://www.eesc.europa.eu/resources/docs/vilgius-panel-3.pdf

Malinska, K. (2004). Organic waste management in agri-food industry in Poland. In M. P. Bernal, et al. (Eds.), *Sustainable Organic Waste Management for Environmental Protection and Food Safety*. pp. 365–368. Retrieved from http://www.ramiran.net/doc04/Proceedings 04/Malinska.pdf

Mars (2012). Mars in Poland. Retrieved November 27, 2013, from http://www.mars.com/global/about-mars/mars-pia/market-summaries/mars-in-poland.aspx

Marthinsen, J., et al. (2012). *Prevention of food waste in restaurants, hotels, canteens and catering*, Copenhagen. Retrieved from http://www.norden.org/en/publications/publikationer/2012-537

Mccabe, K., & Lieberz, S. (2013). *Anti Food Waste Movement Gets Government Support*. Retrieved from http://gain.fas.usda.gov/Recent GAIN Publications/Anti Food Waste Movement Gets Government Support_Berlin_Germany_3-7-2013.pdf

Mihalap, D., & Plepis, A. (2012). *Critical analysis and assessment of actual data on municipal solid waste generation and its processing for all types of waste (Kriticheskij analiz i ocenka fakticheskih dannyh po obrazovaniju tverdyh kommunal'nyh othodov (TKO) i ih pererabotke dlja sovoku*, Minsk. Retrieved from http://www.greenlogic.by/content/files/Othody/Documents/Kriticheskij_analiz_final.pdf

Mihalevich, I. (2013). Separate collection might become mandatory (Razdel nyj sbor mozhet stat' objazatel'nym)'. *Economic Newspaper (Jekonomicheskaja Gazeta)*. Retrieved from http://www.neg.by/publication/2013_09_17_17634.html?print=1

Millers-Dalsjö, D., & Lundborg, H. (2012). *Biowaste slurry collection*. Retrieved from http://www.balticbiogasbus.eu/web/Upload/Supply_of_biogas/Act_4_6/Annex/Biowaste slurry collection.pdf

Ministry of Housing and Utilities of the Republic of Belarus (2013). Regulation on the collecting, storing, recording, transportation and disposal of food waste. Instruction for Use (Porjadok Organizacii, sbora, hranenija, ucheta, transportirovki i unichtozhenija pishhevyh othodov. Instrukcija po primeneniju). Retrieved from http://pinsk.gov.by/economics/instr.doc

Ministry of Natural Resources and Environmental Protection of the Republic of Belarus (2009) Environmental protection and nature use. Waste. Municipal waste management. Objects for solid municipal waste disposal. Rules for designing and operation, pp. 1–29. Retrieved from http://tnpa.by/KartochkaDoc.php?UrlRN=228135.

Ministry of Natural Resources and Environmental Protection of the Republic of Belarus (2013a). *Environmental Protection and Nature Use. Wastes. Rules of Municipal Wastes Management.* Retrieved from http://www.mjkx.gov.by/uploaded/%D0%A2%D0%9A%D0%9F_%D0%BA%D0%BE%D0%BC%D0%BC%D1%83%D0%BD%D0%B0%D0%BB%D1%8C%D0%BD%D1%8B%D0%B5 %D0%BE%D1%82%D1%85%D0%BE%D0%B4%D1%8B_1 %D1%80%D0%B5%D0%B4%D0%B0%D0%BA%D1%86%D0%B8%D1%8F final.pdf

Modin, R. (2011). *Livsmedelssvinn i hushåll och skolor—en kunskapssammanställning (Avoidable food waste in households and schools)*. Retrieved from http://www.slv.se/upload/dokument/rapporter/mat_miljo/2011_livsmedelsverket_4_livsmedelssvinn_i_hushall_och_skolor.pdf

Moora, H. (2011). Municipal waste management in the Baltic States—main trends and emerging issues. Retrieved from http://www.recobaltic21.net/downloads/Public/Conferences/Emerging trends and investment needs in waste management 2011/harri_moora.pdf

Mott, R. (2012). *White paper on household/restaurant food waste*. Retrieved from http://www.ceeres.eu/index.php?option=com_phocadownload&view=category&download=20%253Awhite-paper-on-householdrestaurant-food-waste&id=1%253Ageneral&Itemid=8&lang=en

Multi Communications (2011). Environment-friendly kitchens. Retrieved January 27, 2014, from http://www.multipr.pl/en/cms/news/492,environment-friendly-kitchens/

Order No 14/8a of the Ministry of Natural Resources and Environmental Protection of the Republic of Belarus and the Ministry of Housing and Utilities of the Republic of Belarus (2000). *Order No 14/8a of the Ministry of Natural Resources and Environmental Protection of the Republic of Belarus and the Ministry of Housing and Utilities of the Republic of Belarus of*

19 January 2000 on approval of the regulatory guidance on the selection and, Belarus. Retrieved from http://pravo.levonevsky.org/bazaby09/sbor60/text60337.htm.

Polish Council of Ministers (2010). National waste management plan 2014, pp. 1–83. Retrieved from http://www.mos.gov.pl/artykul/3340_krajowy_plan_gospodarki_odpadami_2014/21693_national_waste_management_plan_2014.html

Polish Food Bank (2012). FOOD Bank in Poland. Retrieved from http://ec.europa.eu/digital-agenda/events/cf/ictpd12/document.cfm?doc_id=21830

Radziszewski, M. (2011, September). Directions of development of biogas plants in Poland by 2020, pp. 1–18. Retrieved from http://www.imp.gda.pl/bkeeold/wydarzenia11_pliki/gaz_forum/szkoleniowa/Radziszewski.pdf

ReFood (2013a). Core Principles and Objectives. Retrieved December 16, 2013, from http://www.refood.eu/en/rfeu/the-company/mission-statement/

ReFood (2013b). Turning yesterday's food waste into tomorrow's energy. Retrieved December 16, 2013, from http://www.refood.eu/en/rfeu/special/home/

Resolution No. 26 of the Ministry of Housing and Utilities of the Republic of Belarus (2003). *Resolution No. 26 of the Ministry of Housing and Utilities of the Republic of Belarus of 30 July 2003 on approval of the instruction on organisation of separate collection, storing and transportation of municipal solid waste (Postanovlenie Ministerstva Zh*, Pinsk: National Register of Legal Acts (Nacional'nyj reestr pravovyh aktov) No 8/9905. Retrieved from http://minpriroda.gov.by/download.php?p_v_id=3927.

Resolution No 6 of the Council of Ministers of the Union State (2010). *Resolution No 6 of the Council of Ministers of the Union State of 23 April 2010 on the Scientific and Technical program of the Union State Improving the efficiency of food production by processing its waste with use of advanced technologies and equipment*, Belarus, Russia.

Resolution No 63 of the Ministry of Natural Resources and Environmental Protection of the Republic of Belarus (2011). *Resolution No 63 of the Ministry of Natural Resources and Environmental Protection of the Republic of Belarus of 31 December 2010 on amendments to the Resolution No 85 of the Ministry of Natural Resources and Environmental Protection of the Republic of Be*, Belarus: National Register of Legal Acts of the Republic of Belarus (Nacional'nyj reestr pravovyh aktov Respubliki Belarus') No 19, 8/23266. Retrieved from http://www.mrik.gov.by/dfiles/000813_205282__na_spetsvodopolzovanie__31.doc.

Rettie, C. (2013). Germans start selling supermarket waste online. *The Food Bankers*. Retrieved November 27, 2013, from http://thefoodbankers.wordpress.com/2013/02/19/germans-start-selling-the-fruits-of-their-dumpster-diving-online/

RREUSE (2010). *Inventory of the policies and stakeholders of waste prevention in Europe*. Retrieved from http://www.ewwr.eu/sites/default/files/Etude RREUSE_EN_20 03 12_4.pdf

Saraiva Schott, A. B., et al. (2013). Potentials for food waste minimization and effects on potential biogas production through anaerobic digestion. *Waste Management and Research*, *31*(8), pp. 811–819. Retrieved December 19, 2013, from http://www.ncbi.nlm.nih.gov/pubmed/23681829

Schneider, F. (2008). *Wasting food—an insistent behaviour*, Alberta, Canada. Retrieved from http://www.ifr.ac.uk/waste/Reports/Wasting Food - An Insistent.pdf

Shestakovskiy, A., & Gnedov, A. (2010). *Electrical and electronic equipment waste management system in Belarus (Sistema obrashhenija s othodami jelektricheskogo i jelektronnogo oborudovanija v Belarusi)*, Minsk, Belarus. Retrieved from http://ecoidea.by/download/13.

Souznoe Veche (2013). How generate income on waste (Kak othody prinesli dohody). *Union State (Sojuznoe gosudarstvo)*. Retrieved from http://www.souzveche.ru/articles/our-union/20077/.

Stare, M., et al. (2013). *Improved food waste—factors on operations (Förbättrade matavfalls—faktorer för verksamheter)*. Retrieved from http://www.smed.se/wp-content/uploads/2013/04/Slutrapport1.pdf

Statistics Division of FAO (2000). *Technical Conversion Factors for Agricultural Commodities*, FAO. Retrieved from http://www.fao.org/fileadmin/templates/ess/documents/methodology/tcf.pdf.

Stenmarck, Å., et al. (2011). *Initiatives on prevention of food waste in the retail and wholesale trades*, Copenhagen. Retrieved from http://www.ivl.se/download/18.7df4c4e812d2da6a416800089028/B1988.pdf

Sundt, P. (2012). Nordic study on avoidable food waste in the hospitality sector, (June), pp. 1–22. Retrieved from http://ostfoldforskning.no/uploads/dokumenter/Food Waste juni 2012/Hospitality sector - PSundt, Mepex Consult AS.pdf

Swedish Environmental Protection Agency (2005). *A Strategy for Sustainable Waste Management*, Stockholm. Retrieved from http://www.naturvardsverket.se/Documents/publikationer/620-1249-5.pdf

Swedish Environmental Protection Agency (2009). *Digested food waste powers Linköping's buses*. Retrieved from http://www.naturvardsverket.se/Documents/publikationer/978-91-620-8401-1.pdf

Swedish National Food Agency (2013) Take care of the food—minimize food waste. Retrieved December 17, 2013, from http://www.slv.se/en-gb/Group1/Food-and-environment/Take-care-of-the-food–minimize-food-waste/

Sysav Biotec (2012). *Annual report 2012*, Malmö. Retrieved from http://www.sysav.se/Global/Informationsmaterial-broschyrer, %C3%A5rsredovisningar, faktablad, rapporter etc/broschyrer och %C3%A5rsredovisningar p%C3%A5 andra spr%C3%A5 k/ANNUAL REPORT 2012.pdf?epslanguage = sv

Sysav Biotec (2013). *From food waste to new resources*, Malmö. Retrieved from http://www.sysav.se/Global/Informationsmaterial-broschyrer, %C3%A5rsredovisningar, faktablad, rapporter etc/broschyrer och %C3%A5rsredovisningar p%C3%A5 andra spr%C3%A5 k/From food waste to new resources.pdf?epslanguage = sv

Trofimovich, A. (2013). Useful waste. What types of waste are processed in Belarus (Poleznyj musor. Kakie othody pererabatyvajut v Belarusi). *Arguments und Facts in Belarus (Argumenty i fakty v Belorussii)*. Retrieved from http://www.aif.by/social/item/21464-musor.html

Tylaite, K., & Bastys, M. (2013). *FoRWaRD Regional Report*. Retrieved from http://foodrecoveryproject.eu/wp-content/uploads/2012/11/Regional_report-Lithuania.pdf

Ujhelyi, K. (2013). *FoRWaRD Survey Report*, Retrieved from http://foodrecoveryproject.eu/wp-content/uploads/2012/11/FoRWaRd-D3.3_Report_of_Analysis_of_Results.pdf.

University of Applied Sciences Münster and Institute for Sustainable Nutrition and Food Production—ISuN (2012). *Reducing Food Waste—Identification of causes and courses of action in North Rhine-Westphalia. Abridged Version*, Münster. Retrieved from https://en.fh-muenster.de/isun/downloads/120613_iSuN_Reducing_food_waste_-_Abridged_Version.pdf

UrbanBiogas (2013). City of Valmiera. Retrieved December 9, 2013, from http://www.urbanbiogas.eu/city-of-valmiera

Veidemane, K., et al. (2004). *Waste management in the Baltic States*, Riga, Latvia. Retrieved from http://www.bef.lv/data/file/waste_management.pdf

Vorne, V. (2012). FOODWEB—Baltic environment, food and health: from habits to awareness. Retrieved from http://foodweb.ut.ee/s2/109_106_26_Feasibility_Study.pdf

Westregård, H. (2012). *Why don't you eat your food? A study about cooperation in the Swedish food value chain to reduce household food waste*. Lund University. Retrieved from http://lup.lub.lu.se/luur/download?func=downloadFile&recordOId=3437018&fileOId=3437020

W-Fuel (2011). Measures for biowaste and sludge prevention. Retrieved November 17, 2013, from http://www.wfuel.info/news.php?id=44

Zhechkov, R., & Viisimaa, M. (2008). *Evaluation of waste policies related to the Landfill Directive*, Copenhagen. Retrieved from http://scp.eionet.europa.eu/publications/wp2008_3/wp/wp2008_3.

Ziedot.lv (2013) For a Fed Latvia. Retrieved December 17, 2013, from http://www.ziedot.lv/en/project/870

Chapter 8
Discussion

The results of the conducted study reveal a number of similarities regarding the state of the problem of food waste in all seven countries, as well as factors, influenced by the national economic, social and waste management situations. Peculiarities of each country have also reflected in the outcomes of the implemented measures, aimed at reduction of food waste.

In each country, the problem of food waste -as a problem- is handled at different extents. Based on this factor, the countries could be classified from the least to the most developed. On the 'least end' is Belarus, on the 'most end' are Germany and Sweden.

One of the main findings of the study relates to data and information availability.

Data and Information

A limited amount of available information and data with regard to the issue of food waste in all countries is one of the main obstacles for the current and other future studies. The analysis of the available sources leads to a number of observations about possible reasons of this situation, such as:

- The lack of data is a result of insufficient level of emphasis given to the problem of food waste in a country. Since, the issue is not one of the government priorities or/and public interests, few related studies or research were/are conducted.
- Information and statistical data exist only to a limited extent. When available, they are mostly found as hard copies, not widely available to the public and used only for internal purposes, by governmental institutions, organisation and companies.
- Data and information related to the topic, including taken measures, activities, and legislation are often not available in English or in Russian. It might be an obstacle for possible future research or studies from outside of a country, especially, if there are reasons not to conduct them, inside of a country. Furthermore, such situation significantly limits possibilities for sharing and/or exchange of knowledge and experience among countries.

W. Leal Filho and M. Kovaleva, *Food Waste and Sustainable Food Waste Management in the Baltic Sea Region*, Environmental Science and Engineering, DOI 10.1007/978-3-319-10906-0_8

Moreover, the relatively low level of emphasis given to the subject matter of food waste, means that this topic is not as high in the political or scientific agendas as it should.

Food waste classification

The analysis of the available sources shows that there is no single definition of the term 'food waste'. In different countries, or even in different studies related to the same country the term might include different food related types of waste. In addition, as a sub-category, food waste is classified under different wider categories, not only across the countries, but by different institutions, in related reports and legislative documents, in a single country. Such situations obstruct not only data collection and reporting, but implementation of the measures aimed at reduction of food waste as well.

Causes of food waste generation

The findings show that, mostly, the causes of food waste generation are similar across the discussed countries, as well as, to those, identified for other developed countries. However, it is important to note that the extent of the same causes is different. For example, low food prices and consumer behaviour are one of the key causes of food waste generation in Germany, whereas, Belarus works on over-coming challenges related to the lack of technologies and facilities for processing of by-products that, otherwise, are discarded.

Another cause of food waste, packaging, to some extent, together with marketing instruments, could be named as a tool that retailers use to 'transfer' potential food waste from supermarkets to households. The issue raises a number of controversial discussions. On the one hand, it 'enforces' a consumer to buy unwanted amounts of food products, on the other hand, a bigger variety of packaging sizes, which might potentially reduce the amount of food waste, and definitely will increase the amount of packaging. Moreover, more opportunities to buy more products in bulk require more resources, e.g. higher energy consumptions, and/or labour, as well as aggravate a problem of storage and faster spoilage of those products.

8.1 Food Waste Generation

The findings of the study show that there is no clear negative correlation between the amount of food waste accumulated and the share of consumer expenditures on food in a country. Based on the analysis of the data on consumer expenditures, the countries could be classified in the following order (from the smallest share of food expenditures to the biggest, in 2012):

1. Germany,
2. Sweden,
3. Latvia,

4. Poland,
5. Estonia,
6. Lithuania,
7. Belarus

Whereas, according to the results of the BIO Intelligence Service study, the sequence (from the biggest to the smallest amount of food wastage generated) is slightly different:

1. Germany,
2. Poland,
3. Sweden,
4. Lithuania,
5. Estonia,
6. Latvia

However, this research did not include Belarus. Thus, already, at this stage, the comparison does not show a strict dependence between these two factors. The same pattern is observed in the results of the comparison of the share of food expenditures and the amount of household food waste per capita. The amount is calculated based on both, the results of the BIO Intelligence Service study and the assumption that the share of household food waste in household food wastage accounts to 25 % in all countries. Thereby, the statement that those households, who spend less, throw more food is not entirely correct. To some extent, this argument is valid only for Germany and Sweden. For the rest of the countries the correlation is positive. It means that, at the same time, consumers spend and waste more.

The analysis of ratios of the expenditure share, amounts per capita of food wastage and of food waste shows that there is also no proportionality between these parameters (Table 8.1).

For instance, German households spend on food 1.72 times less than Latvian households but generate 2.5 times more food waste (Table 8.1). In other words, the fact that a consumer, for example, spends on food twice more, does not imply that an amount of generated food waste will be also twice bigger.

Table 8.1 Ratios of the expenditure share, amount of food wastage and of food waste (own calculations)

	Germany	Sweden	Latvia	Poland	Estonia	Lithuania
Expenditure share ratio	1.72	1.54	1.0	0.96	0.96	0.73
Ratio of the amount of food wastage per capita	1.2	2.2	1.0	2.3	2.6	1.8
Ratio of the amount of household food waste per capita	2.5	2.6	1.0	1.5	1.7	0.95

The obtained results call into question the role of low or even very low food prices as one of the main causes of food waste. The issue could be possibly true, but only for a limited number of countries, for example, Germany and Sweden. Moreover, this fact must be taken into account, when recommendation regarding measures, aimed at reduction of food waste, are made.

Another factor indicated by many studies, not only in the discussed here countries, that strongly influences on the amount of food waste generated are consumer behaviour. The analysis of the results of the Eurobarometer study about households' perception of amounts of food they throw, together with the results of the BIO Intelligence Service, discussed above, reveals curious dependence. In the countries, where households generate the biggest amounts of food waste, the majority of respondents indicated that they throw less than 15 % of food, whereas, the average value across the Europe is 25 %. However, it is important to note that in all six countries people admit that they discard food. Thus, the main question is about their perception of an amount of food that they send to bin. These findings, one more time, point out to consumer behaviour as to one of the main causes of household food waste.

8.1.1 Food Waste Amounts According to the FAO Food Balance Sheets

FAO is one of the few organisations that provide comparable statistical data regarding food wastage in each of the countries discussed here. The calculation and the later analysis of the results of the amounts of food wastage, generated per one hundred tonnes of total available supply for different categories of the products partly put into question the commonly held opinion that in more developed economies an amount of food wastage generated during manufacture, storage and transportation stages is much less than in less developed economies. For example, Germany, Sweden and Poland generate more vegetable, potato and cereal wastage than Latvia, Estonia or Belarus. However, based on the available data it is impossible to determine the exact stage, where, the generated amount of wastage is the biggest. Since, the FAO waste data also include amounts of wastage that are a result of the imbalances of supply and demand, it would be interesting to see, what are the main causes of food waste, at each of these stages, either problems relate to infrastructure and facilities, or to poor management (bad planning, wrong ordering, poor forecasting of consumer demand and etc.).

The calculation and the later analysis of the results of food waste of each food category show that Poland, Germany and Sweden generate more vegetable, cereal and potato waste as well (Table 7.3).

8.1.2 Food Waste Generated Based on the FAO Technical Conversion Factors—Extraction Rates

The amount of food waste accumulated from the processing of primary products point to a number of aspects. Firstly, the highest values of extraction rates and the amounts of generated food waste, vary among the countries. Secondly, there is no one country with the highest values for all types of products. It could be concluded that an economic situation in a country not always an evidence of use of best available technologies during the food manufacturing process. For example, Sweden and Poland generate the biggest amounts of waste during the production of bran of wheat, Lithuania and Sweden, during the production of malt of barley, and Poland and Latvia, during the production of bran of rye. At the same time, Sweden and Germany are 'leaders' in generating molasses waste. Together with Belarus, Sweden also generates the biggest amounts of waste from the production of oil from sunflower seeds. Considering that, the obtained results, calculated based on the fact that at least one of the seven counties produces a product with using of the best available technologies, could be interpreted as follows. One of the possible reasons of no introduction of those technologies into the production process in the rest of the countries is financial. Companies unable or unwilling to make this kind of investments because, for example, costs of managing generated waste are relatively low and do not have a strong effect on a company's profit and, therefore, cannot be seen as an incentive to reduce this type of waste.

8.1.3 Situation in Individual Countries

8.1.3.1 Belarus

Since the country is not a member of the EU, none of the European food wastage/ waste legislation or related studies cover this country. The problem of food waste is not well investigated in the country. Therefore, there are little studies or research in regard with the issue, which also mean a lack of statistical data. Despite these facts, Belarus is the only country, where food waste is, to some extent, regulated by the national legislation. However, it does not mean that its enforcement has the same effect as campaigns and other activities taken place in other countries. It is important to mention that the regulations are not applied to households. Thus, it is possible to assume that data regarding food waste exist, but, for some reasons, are not available to the public. This assumption could be also supported by the fact that the Ministry of Natural Resources and Environmental Protection of the Republic of Belarus has not responded to the request to provide data or at least information about food waste, for the purpose of this study.

Based on available information, mostly in Russian, the findings show that the main problems regarding food waste is the problem of by-products generated by the food industry. It is important to say that the problem is defined as a problem, openly discussed and there are a number of measures, taken at the governmental level. The main cause of this type of food waste is lack of or obsolete technologies for treating generated amounts of by-products.

It is very difficult to say something about food waste generated by the retailer sector and households. The obligation to report about amounts of discarded food and reasons for these might lead even to illegal actions taken by stores. Usually, such evidences are discussed in the local media.

According to the official sources, the share of household food wastage equals to 27 %, however, it is impossible to define, the food waste fraction in the total amount. The fact that more than 91 % of population get food products from their or their relatives' subsidiary plots, not only indicates the food value, but also has a great impact on person behaviour. Usually, it makes much more difficult to discard something, when a lot of personal efforts were put into growing it.

8.1.3.2 Estonia

The study shows that information and data regarding food waste, generated in the country, are very scarce. The data about kitchen waste are available only for the year 2005 and today, it could be considered as not relevant.

The main sources are the studies, conducted at the European level, such as the BIO Intelligence Service study. According to the study, the manufacturing sector generates the biggest amount of food wastage in comparison to other sectors in the country. It might be a result of obsolete equipment and/or lack of facilities to process by-products. The share of food wastage is one of the smallest among the discussed countries. This fact gives a reason as to why some authors claim that the country has had less problems with food wastage generation. However, this statement is possibly wrong because of lack of data.

The amount of household food waste, calculated based on the BIO Intelligence Service study is also relatively small. Furthermore, according to the Eurobarometer study, the highest percentage of respondents, who claimed that they do not throw food, was in Estonia, but again, there are no food waste statistical data to support or refute these results.

The only organisation that is active in the field of food waste in the country is the Estonian Food Bank.

The causes of food waste generation were named the same, as in other countries, such as insufficient information or knowledge, weak legislation and government reluctance to the problem.

8.1.3.3 Germany

In comparison to other discussed countries, the number of studies regarding the problem of food waste in the country conducted at the national level is one of the highest. At the same time, Germany generates the biggest quantities of food waste. However, there are still problems with the classification of waste types that could be referred as food waste. Most of the experts and researches use the WRAP classification, according to which, food waste could be one of the three types, avoidable, partly avoidable, or unavoidable. However, such types as food residues, food losses, and kitchen waste in different occasions could be referred to all these types. It creates a number of obstacles during the comparison of the results obtained in the different studies.

Nevertheless, the findings at the national and European levels indicate that German households generate the biggest amounts of food waste in comparison to other sectors. The main difference in results is in regard with the share of food waste in the total amount of household food wastage. In contrast to 25 %, the commonly acceptable average value for the EU countries, the German researches claim that the value ranges between 47 and 65 %. Therefore, such cause of food waste as undervalue of food, mostly, because of low food prices could be considered as most relevant. Although, other causes such as lack of incentives to prevent or at least reduce an amount of thrown food, insufficient information, knowledge and education about environmental consequences, efforts and resources, required to produce food products, as well as, a notion of their infinite availability also require close attention.

Among the main categories of food products that constitute food waste are fruit, vegetables and bakery products. This fact, partly put into question 'confusion of date labels,' as one of the major causes of food waste, because the products mentioned above mostly come without date labels. It means that households discard these products only based on their own judgment and not on information on packages. At the same time, it makes the problem of lack of knowledge about how to store products properly and buying too much more prominent.

The largest amounts of household food waste do not make the extent of the problem in other sectors less noticeable. According to a number of studies, almost one third of food waste generated in the food industry is a result of technical faults. Together with food products discarded due to the requirement of quality management, it comprises more than 60 % of the amount of food waste generated by the industry. These findings, one more time, raise a question about efforts that company make to eliminate or at least to reduce their food waste, even, if they classify it as losses, residues or by-products. Another question is if companies have enough financial incentives to change the situation, in other words, if an effect from reduction of costs for treatment of the generated amount of food waste on a company's profit is significant enough in comparison to the costs of required measures to treat this waste.

8.1.3.4 Latvia

Like in Estonia, the data regarding food related waste is available for the years 2004–2006. The only recent data about the situation of food wastage/waste in the country come from the international studies. It is also possible that such information and/or data are available but only in Latvian language. The analysis of media resources shows that occasionally local journalists raise the issue as well.

According to the BIO Intelligence Service study the manufacturing sector generates the biggest amount of food wastage in comparison to other sectors in the country. Like in Estonia, it might be a result of obsolete equipment and/or lack of facilities to process by-products. The share of food wastage, as well as, the share of household food waste are smallest among the discussed here countries. According to the media sources, it might be explained by cost-conscious and food respectful consumer behaviour. It is possible to assume that the problem of food waste will become more discussable in the country in the coming years, first of all because of the continuous discussion of the issue at the EU level.

8.1.3.5 Lithuania

Most of the available data on food related waste are referred to waste generated by the meat, fish, dairy and bakery industries, with no reference to a possible percentage share of food waste. The annual increase in the amounts of generated wastage could be associated with, for example, an increase of production, whereas, a wastage share per one tonne stays constant. However, in any given year the amount of food wastage, generated by the dairy industry, is the biggest.

The causes of food waste in retail chains, stated by their representatives, are the same as were identified in the mentioned above studies, such as poor planning and consumer requirements for particular product standards. The problem of ineffective stock management could be considered as an 'inherited' problem, because many of local chains are branches of the companies that operate in other countries as well. Therefore, it is hard to believe that if a chain has such problem in one country, it would be solved in another country, especially, where the problem of food waste is not well developed and regulated.

During the study, conducted by the Lithuanian Food Bank, representatives of companies listed a number of causes of food waste. This includes: (Tylaite and Bastys 2013)

- Mistakes in planning the demand;
- Mistakes in planning the sales volume;
- During the selling process, when buyer changes his/her mind and leave the short term product anywhere in the shop;
- Insufficient information and knowledge about possible ways of managing food waste;
- Lack of information about (charitable) organisations that work with food waste, tax regulations on donating food, regulations on food safety for food waste donations.

But perhaps the most important point is not the mentioned causes, but the fact that they knew these causes. It means that despite the lack of an open discussion of the problem by the industry, companies are aware of it and, even, conducted inner studies to identify the causes. It is also important to note that companies' expectations to get something for food donation e.g. financial incentives, diminish or even eliminate a possibility of voluntary cooperation that brings back to the issue of the governmental role in the problem of food waste prevention/reduction.

8.1.3.6 Poland

The country is another 'leader' in food wastage and waste generation. However, the pattern according to which the problem is developing is entirely different in comparison to other countries. Moreover, some of the obtained results contradict to the general results in other discussed countries, particularly with regards to consumer behaviour, knowledge and effectiveness of applied measures.

According to the available studies, the food industry generates the biggest share of the total amount of food wastage, as a result of obsolete technologies and facilities. As well as in Lithuania, it is almost impossible to identify the share of generated food waste. Of course, it could be expected that improvement of management practices and modernisation of the technical potential would reduce an amount of waste generated. However, there are two main obstacles, the first one is potential costs, the second one is willingness of companies to make these improvements.

A number of studies, including the study of the Polish Food Bank, conducted in regard with causes of food waste in the country, indicated the same points, as in other countries. It also was pointed out that there is dependence between an amount of generated food waste and material status. These findings reveal a huge social and economic gap in the country, when, on one side, there are millions of people, who live in deep poverty and on the other, a large number of people, who throw food because can afford it.

Results of another study contradict the experts' opinion and findings in other countries that lack of information and knowledge about social, economic and environmental consequences of food waste are its main causes. Many consumers in Poland are aware of all of these. Based on these facts, it could be said that people throw food not only because they can afford it, but also because they do not care and it is not important, in other words, this situation is an example of almost total indifference to the problem. Therefore, majority of measures aimed at reduction of food waste that are recommended for other countries cannot be implemented in Poland.

8.1.3.7 Sweden

The type of food related waste studiesmostly differs from those in other countries. First of all, researches were interested in the problem, already at the end of 70th. Secondly, until today, most of the conducted studies are very specific: every time focusing on a particular institution. However, considering the fact that Swedish municipalities differ in a wide range of aspects, including the waste issues, it makes extrapolation of the results i.e. their generalisation at the country level very difficult. Nevertheless, the narrowly focused studies could be considered as representative case studies for different sectors, not only in Sweden, but in other countries with a similar state of economy.

The value of a percentage share of household food waste varies in different sources; however, the most acceptable is 35 %. It is stated and used by governmental agencies. On the one hand, it is much smaller than in Germany, but on the other hand, it is still bigger than the value set at the European level.

It is important to note that most of the available studies are focused on food waste generated by all sectors except households. The possible reasons could be the following:

- The share of food waste in household food wastage is significantly smaller than in other sectors;
- The studies regarding household food waste are much more difficult to conduct;
- Implementation of measures aimed at reduction of food waste is easier, for example, improvement of management practices, as it was realised in a number of schools.

The value of the percentage share of food waste generated by other sectors varies at large extent. Whereas, some studies bring out such values as 18 % for schools, approximately 67 % for the hospitality sector, and 20 % for food service institutions. According to other studies, the share of food waste in food wastage discarded as mixed municipal waste is the following: restaurants—about 62 %, food stores—91 %, and schools—52 %. These results, once more,indicate thathouseholds are possibly not the main producers of food waste, thus making the dependence between food prices and an amount of generated waste even weaker.

The types of discarded food products are, mostly, the same as in other countries, e.g. fruit, vegetable, bakery and dairy products. The same is applied to the causes of food waste generated in the retail and wholesale sectors, e.g. passed best before/expiry date, oversupply, improper handling of products and challenges to predict consumers demand. The studies also indicate that there is no difference between types of food discarded by small and large store. Though, initially, it seems more obvious to assume that differences in management practices, planning, available labour force and facilities, as well as in methods of monitoring and control, significantly influence on an amount of generated food waste.

8.2 Food Waste Treatment

8.2.1 Belarus

The analysis of the waste treatment methods show that the current situation is similar, at large extent, to the situation in Poland, Latvia, Lithuania and Estonia before implementation of the Waste and Landfill Directives. Therefore, it is possible to assume that in the near future, most of the measures will focus on treatment of bio-waste. Thus, food waste will also be treated in the same way, i.e. by methods from the lower part of the food waste hierarchy. However, currently separate collection required for implementation of these types of treatment only applied in a few companies.

The study also shows that for households and companies, the financial aspect has a key role in all waste related issues. The long tradition of very low waste tariffs that comes from the Soviet times prevents effective implementation of required measures, including building of infrastructure. The results of the study in one of the regions showed that most people, regardless of age and type of housing, support the idea of waste sorting. However, majority of them are not ready to pay more in order to improve quality of waste collection and disposal (e.g. regular waste collection, better waste containers, organization of landfills that eliminates a threats to human health) (Executive committee of Kobrin (Belarus) 2012). It makes the problem more complicated, because experts talk about a necessity to increase tariffs of MSW disposal (Mihalap and Plepis 2012).

The country, to some extent solves the problem by cooperating with international organisations that also help them to introduce new technologies of waste treatment, for example, distribution of bio-composters for home composting on a grant basis.

As a very strong regulator, the government has a lot of means to 'enforce' companies and organisations not only to implement separate collection, but also, to take measures regarding food wastage and food waste treatment. The main point, here, is willingness of the government to do so. Until today, as the research shows, most of the measures were reflected only in the legislative documents. One of the practical examples is the initiation of the project, aimed at reduction of the amount of by-products, generated by the food industry. However, it is possible to assume that the main reason was economical and not ethical or environmental, because possible financial losses from non-treatment of this type of waste were so significant that motivated the government to take a number of steps.

8.2.2 Estonia

There seem to be no public awareness or/and education campaigns regarding the issue of food waste, at least available in English or Russian. However, there are a number of measures, including this type of activities, used to promote and stimulate sustainable distribution and consumption. Therefore, it is possible to assume that with slight changes, the issue of food waste could be easily added to these campaigns.

Today, the Estonian Food Bank is the key actor in the field of food waste treatment in the country. Therefore, the main method from the upper part of the food waste hierarchy applied for food waste reduction is donation. However, the process is limited by the current legislation that indicates on the necessity of the government participation in the process.

Companies that participate in the process of food waste prevention/reduction are financially motivated, in other words, through these practices companies reduce their costs for waste management. Therefore, it is possible to assume that additional financial incentives e.g. tax cuts, might intensify the process of food waste reduction. This fact, one more time points out to the importance of government involvement.

According to the EU regulation, bio-waste should be diverted from the landfills and treated through other methods applied in the country; therefore, it is obvious that food waste is treated as other fractions of bio-waste. It makes to consider that this type of environmental consequences of food waste (e.g. GHG emissions) is or will be solved, orat least reduced.

8.2.3 Germany

The country is an example of how, as main producers of food waste, households, become a main driver for its reduction and prevention. The analysis of the treatment practices also indicates an important role of the federal German government in the process. A large variety of public awareness campaigns, activities and educational programs are considered to be effective in the process of food waste prevention and reduction. However, despite some results, published by the BMELV about effectiveness of the applied measures, it is still hard to say, at what extent, these activities will help to reach the target of halving the amount of generated food waste by 2020. Therefore, additional studies and permanent monitoring of the results are required.

Such government interest and involvement in the process also have a positive effect at the European level. Considering the role of the country in the EU, the problem of food waste could be developed, at even larger extent, than today, including the legislative aspect. The Commission might use the same framework and principles, as were applied for the issue of biodegradable waste.

The analysis of the available studies show that, despite a large number of measures, that cover the largest share of German population, there is almost no initiatives and/or projects, aimed at reduction of food waste, implemented locally, i. e. in a particular institution or company. One of possible justification might be an argument that the share of food waste generated by other sectors is insignificant in comparison to the amounts generated by households.

The practical measures of reduction of food waste generated by the food industry, are, mostly, implemented by food banks. The German network of the Food Banks could be considered as one of the biggest and oldest among the food banks in the discussed countries. However, in case of Germany, such donation practices might also have drawbacks. For example, donations do not motivate companies prevent or reduce an amount of potential food waste that they generate, since the easier and cheaper solution is to transfer oversupply and/or overproduction to the Food Banks.

The treatment methods of food waste from the lower part of the food waste hierarchy are also well developed and established in the country. Therefore, in principle there are no problems with diminishing some of environmental consequences of food waste, e.g. GHG emissions.

Mainly, the problem of food waste in Germany is related to the social and ethical issues, as well as to the problem of sustainable use of resources.

8.2.4 Latvia

As well as in Estonia, there is no available information in English or in Russian on public awareness or/and education campaigns regarding the issue of food waste. However, there are a number of activities focused on promotion of sustainable distribution and consumption. Therefore, in the same way, as in Estonia, it is possible to assume that, with slight changes, the issue of food waste could be easily added to this type of campaigns.

The main method of food waste reduction from the upper part of the food waste hierarchy is donation of food to the Latvian Food Bank. Retailers also implement a number of measures, in order to reduce their food waste; however, there is no data that make it possible to calculate the amount of food waste that is prevented through these methods.

The issue of food waste treatment is not seen to be one of the prior issues in the area of waste management. The country is still struggling with implementation of separate bio-waste collection, particularly, in cities. It is possible to assume that household food waste is or will be treated together with other types of bio-waste.

Companies are and will be financially motivated to reduce an amount of generated food waste, because of the required contracts for waste disposal. The process will become more intensive from the moment, when local governments will start actual control of the binding regulations regarding companies' cooperation with organic waste service companies. However, the more important question is how these companies will treat collected waste. Because, according to some studies today, there are no data available about volumes and means of utilizations of these waste streams by this type of companies.

8.2.5 Lithuania

As well as in Estonia and Latvia, there are public awareness or/and education campaigns related to the problem of sustainable distribution and consumption that possibly could also include the issue of food waste. The only campaign aimed at reduction of food waste was organised by the Lithuanian Food Bank.

Donation of food to the Food Bank is the main method of food waste reduction/prevention from the upper part of the food waste hierarchy. The cooperation of the organisation with one of the retail chains shows willingness of companies to take steps in order to reduce their food waste. The fact that this retailer is presented on the Latvian market opens new possibilities of applying similar practices in Latvia as well.

One of the main problems of food wastage treatment is an effective treatment of by-products. Unlike, for example, in Belarus, where there is the lack of required technologies and facilities, Lithuania has available technical capacities for this type of treatment. However, there are evidences of ineffective use of those, which, among other things, indicate on an insufficient number of incentives for their application by companies.

It is possible to assume that the government will be more active in this area, since the country has to reach the EU waste targets and, today, this waste stream goes to landfills.

The fact that companies of the food industry, at particular extent, have to confront the problem of treatment of generated waste alone indicates poor participation of the key actors, for example, government participation.

8.2.6 Poland

The main obstacle, but at the same time one of the advantages, is the fact that the country is still working on reaching the targets set by the Waste and Landfill Directives. It means that there are possibilities to include the issue of food waste in all activities focused on the implementation of the EU waste requirements.

Currently, donation of is one of the major methods of prevention/reduction of food waste generated by the food industry. As well as the German Food Banks, the Polish Food Bank is one of the oldest of this type of organisations in the discussed countries. The Federation of Food Banks is also very active in initiating and supporting different types of public awareness, education campaigns and related projects. In could be said that, to some extent, the role of the Federation in the area is similar to that, played by the BMLEV in Germany. There is an established cooperation between the industry and the Polish Food Banks as well.

The analysis of food waste treatment methods reveals a paradoxical situation, when on the one hand, the country has a very long history in utilisation of by-products, but on the other hand, the manufacturing sector generates the biggest share of food wastage. There are two possible reasons for such situation. One reason might be ineffectiveness of the applied practices. Another reason is that available data are not reliable, because classification of food waste/wastage strongly differs from that in other countries and amounts of utilised by-products are included in waste statistics as one of the waste categories.

8.2.7 Sweden

The country represents an alternative version of food waste treatment. For example, there are almost no widely spread awareness-raising food waste reduction campaigns. However, the issue is included in the waste management plan. The projects, aimed at reduction of food waste, are implemented successfully, but locally, in other words, in a particular organization or municipality. It makes difficult to claim that, later, the same practices will be implemented in that type of institutions or in other municipalities across the country. Sweden is also an example of cooperation between different types of organisations and institutions, whereas, for example, in Germany, the BMELV only calls organisations upon similar cooperation. Unlike in other countries, there are a big numbers of published reports and guidelines regarding food waste reduction activities, however, the more important question is how organisations get informed about these publications.

In contrast to other countries, the issue of food banks is not developed. It seems that companies prefer to eliminate this intermediary and treat their food waste individually.

Based on the mentioned above evidences it could be concluded that the country has implemented enough measures to get successful results, but the problem still remains.

The waste treatment methods from the lower part of the waste management hierarchy are strongly developed and implemented as well. The analysis of available studies gives impression that food wastage is the only type of waste that is still not treated with the same efficiency rate as other types. Today, infrastructure and facilities for food wastage treatment are so easy available and reachable for households and companies that it, in some way, demotivates them to reduce an amount of generated food waste. Moreover, the country intensively promotes and strongly support the transformation of food wastage into fuels, because one of targets is to raise proportion of energy from renewable sources in gross final consumption of energy from 40 % in 2005 to 49 % in 2020 (Corvellec 2012). It also talks about positive impacts of producing biogas from food wastage as reduction of dependence of the Swedish society on finite resources. In addition there are statements about an increase of people's wellbeing, when they know that food wastage is reused effectively (Eco-Innovation Observatory 2013), with that, the share of food waste in the total amount is unknown.

Furthermore, some retailers prefer send their food waste directly to treatment plants, instead of trying to prevent or reduce it. The issue is taken very serious, for example, according to the inner policy of one of the retail chains, stores are not allowed to give away any type of food products. Therefore food waste (including fruits and vegetables) is sent for disposal, depending on the local infrastructure to a composting facility or to incineration (Eriksson et al. 2012).

However, already today, the country has a problem of overcapacities. To some extent, the fact that country relies on food wastage, as on a source of energy and biofuel, prevents an application of any treatment methods from the upper part of the food waste hierarchy. One of the reasons is that a reduced amount of food waste in the total amount of food wastage enforces companies to look for an equivalent replacement, the main question is how easy and fast they would be able to do it. Of course, it might be assumed that, despite the decrease of the share of food waste, the total amount of food wastage would remain the same, however, this situation is only possible by a drastic increase of food consumption and as a result, of amounts of other fractions of food wastage. To avoid these problems and, at the same time, prevent/reduce the amount food waste, companies might increase import of waste from other countries. However, with years, it might be a problem because other countries also continue to build up their capacities for biological treatment of waste.

The various extents at which the problem is investigated in each country; composition and causes of generated food waste, as well as applied measures, reveal a large number of areas that are still required to be explored. The obtained results indicate on a necessity of cooperation between all actors involved, directly or indirectly, in generation of food waste. These joined activities would have maximum potential and effect in tackling the problem on both, at the country and at the European level.

References

Corvellec, H. (2012). *Normalising excess: An ambivalent take on the recycling of food waste into biogas*, Helsingborg. Retrieved from http://www.ism.lu.se/fileadmin/files/rs/wp/WP_15_NOV_2012.pdf.

Eco-Innovation Observatory (2013). Pre-treatment plant for food waste. Retrieved from November 17, 2013 http://www.eco-innovation.eu/index.php?option=com_content&view=article&id=627:sysav&catid=75:sweden.

Eriksson, M., Strid, I. & Hansson, P.-A. (2012). Food losses in six Swedish retail stores: Wastage of fruit and vegetables in relation to quantities delivered. *Resources, Conservation and Recycling*, *68*, 14–20. Retrieved from http://dx.doi.org/10.1016/j.resconrec.2012.08.001.

Mihalap, D. & Plepis, A. (2012). *Critical analysis and assessment of actual data on municipal solid waste generation and its processing for all types of waste (Kriticheskij analiz i ocenka fakticheskih dannyh po obrazovaniju tverdyh kommunal'nyh othodov (TKO) i ih pererabotke dlja sovoku*, Minsk. Retrieved from http://www.greenlogic.by/content/files/Othody/Documents/Kriticheskij_analiz_final.pdf.

Tylaite, K. & Bastys, M. (2013). *FoRWaRD Regional report*. Retrieved from http://foodrecoveryproject.eu/wp-content/uploads/2012/11/Regional_report-Lithuania.pdf.

Chapter 9
Conclusions and Recommendations

9.1 Conclusions

The current study centres on today's situation and future trends in food waste management in seven Baltic Region countries such as Belarus, Estonia, Germania, Latvia, Lithuania, Poland and Sweden. The findings are based on an analysis of the information and statistical data on the problem of food waste, available from various sources. It also includes such aspects as the causes of food waste, methods of its treatment, best practice, countries' economic situations, with a focus on consumer purchasing power, a food consumption pattern, undernourishment and poverty level, as well as renewable energy production and a state of biodegradable waste management. The obtained results give ground to a number of conclusions.

There is no single, uniform definition of the term 'food waste', not only among the countries, but even in different studies related to the same country. Therefore, the classification of such food related waste types, as food residues, food losses, kitchen waste and etc. is left at the discretion of those who conduct a study.

The very limited number of studies leads to the lack of statistical data, challenges to define stages of the food supply chain, where the biggest amounts of food waste are generated, as well as measures, required to be implemented, to tackle the problem.

In each country, the problem of food waste is tackled to varying degrees.

Across the countries, the role of the main actors differs, for example, in Germany is the Federal Ministry of Food Agriculture and Consumer Protection, in Poland, the Federation of Polish Food Banks, whereas, in Sweden the network 'SaMMa', a platform for cooperation of 30 organisations that represent various sectors.

The percentage share of household food waste, especially, in countries, where more data are available, strongly deviates from the value of 25 %, stated in the WRAP studies.

The share of food waste, generated by other sectors, is largely unknown, particularly, in the manufacturing sector. The majority of studies were undertaken in Sweden and Germany and focused on food waste generated by retailers, schools and restaurants.

W. Leal Filho and M. Kovaleva, *Food Waste and Sustainable Food Waste Management in the Baltic Sea Region*, Environmental Science and Engineering, DOI 10.1007/978-3-319-10906-0_9

Among the food products that are discarded, mostly are fruit, vegetable, dairy and bakery products, and less meat.

The analysis of the fractions of food waste, calculated based on the FAO extraction rates, shows that, to some extent, an economy rating of a country does not have an impact on the amount of food waste, generated as a result of processing of, in most cases primary products into secondary products (e.g. flour from wheat). Thus, Sweden and Germany generate highest amounts of food waste for some categories of products, in comparison to the rest of the discussed countries.

There seems to be no dependence between an economy rating of a country and the amount of food waste, generated per one hundred tonnes of total available supply for different categories of the products, during all stages of the food supply chain, except the pre-harvest, harvesting and household stages. In other words, Germany and Sweden do not always generate the smallest amounts of food waste. However, it should be noted that the obtained results make it impossible to define the exact stage where, most of waste is accumulated.

The results of the study do not strongly support the hypothesis of direct dependence between the amount of accumulated food waste and the share of consumer expenditures on food in a country. The statement is true, mostly, for Germany and Sweden, however, refuted in comparison between Poland, Latvia, Lithuania and Estonia.

The economy ranking is better reflected in varying degrees, in which, the problem of food waste is discussed and treated in each country.

The choice of food waste treatment measures, depends on a number of factors, such as a current waste management situation, including available infrastructure and facilities, the implementation of biodegradable waste separate collection, applied treatment methods, as well as a degree to which, the public is aware of the problem of food waste in a country.

Below, there is a list of the main outcomes of the study, based on the available sources in regards with the state-of-the-art of the problem of food waste in each country.

9.1.1 Belarus

The subject of food waste is not well-developed in the country. Therefore, there is a lack of studies and public activities related to the problem. However, Belarus is the only country, where, in legislative documents, food related wastes are detailed categorised, with categories that could be unified under the type ‘food waste’. Moreover, it is legally binding for companies and organisation to report about amounts of generated food related wastes and reasons of disposal.

The process of separate collection of biodegradable waste is poorly developed and implemented only in a small number of companies and on a small scale in different regions of the country.

The latest available technologies for biodegradable waste treatment, Belarus gets on grant base, through the cooperation with international organisations.

Together with Russia, the country has implemented a project, aimed at reduction of an amount of food waste in a form of by-products, generated by the food industry. The main goal was the development of technologies that allow the future use of accumulated by-products. The project is an example of one of the ways of food waste treatment, when the government is willing to participate in the process.

There are no data on losses during pre- and harvesting stages that could be referred as food waste. It is possible that waste, accumulated during these stages is used as animal feed or left on the fields as a fertiliser.

9.1.2 Estonia

Majority of the Estonian public is not aware of the extent of the problem. Until today, the government involvement in the issue has been very limited. Therefore, the main source of statistical data on food waste is studies, undertaken on the European level.

The results show that the manufacturing sector generates the biggest amount of food waste in comparison to other sectors.

At the same time, the lack of the public interest to the problem might be explained by the facts that according to a number of studies, the country generates one of the smallest amounts of food wastage and households food waste, and the larger share of respondents, in comparison to other countries, states that they do not throw food at all.

There is no available information on applied measures from the upper part of the food waste hierarchy, at least in English or Russian, except on food donation to the Estonian Food Bank. The organisation is the key player in the field of food waste in the country.

The achievements in reaching the EU waste targets and, even, the stricter national targets, make it evident that food waste will be treated, at least, in the same way, as other fractions of bio-waste. Thus, the main treatment methods of food waste applied in the country are methods from the lower part of the food waste hierarchy.

9.1.3 Germany

The country is one of the 'leaders' in the field of food waste, in both generated amounts and applied measures, particularly from the upper part of the food waste hierarchy.

The most of food waste is generated by households. The share of discarded food almost two times exceeds the value (25 %), indicated in the WRAP studies.

Among the main causes of food waste are low food prices, lack of public awareness about consequences of the problem, insufficient knowledge on food storage, date labels, and usage of leftovers, underestimation by households of the amount of food they throw away, as well as poor planning while grocery shopping.

There is very strong governmental involvement from the part of the BMLEV. The Ministry initiates and supports a large variety of projects, campaigns and initiatives, aimed at increasing public awareness and knowledge regarding the problem of food waste.

The German network of Food Banks is one of the oldest in the discussed countries. As a result, donation is another important method of reduction of food waste, which comes from producers, wholesalers, retailers and other stores.

The methods of bio-waste treatment from the lower part of the waste hierarchy are developed as well. Therefore, discarded food is also treated together with other fractions of bio-waste. Among the prevailing treatment methods are anaerobic digestion, composting and MBT.

9.1.4 Latvia

The statistical data on food waste come, mainly, from the European studies. According to which, the country generates the smallest amounts of food wastage and households food waste.

The manufacturing sector generates the largest amount of food wastage; however, there are no data on the fraction of food waste in that amount.

Based on the available sources, it is very challenging to identify the degree to which the Latvian public is aware and interested in the problem. The only evidence is an occasional publication on the issue in the local media.

There is no available information on applied measures from the upper part of the food waste hierarchy, at least in English or Russian, except on food donation to the Latvian Food Bank. The issue of food waste might be possibly added to a number of campaigns, focused on promotion of sustainable distribution and consumption that are launched in the country.

Retailers do not provide data on the amount and treatments methods of their food waste. The only known fact is that companies are required to have a contract for waste disposal with companies that provide this type of services, however, amounts of waste and methods of its treatment are also unknown.

The country is still struggling with the implementation of separate bio-waste collection, particularly, in cities. Supposedly, food waste is treated in the same way as other fractions of bio-waste.

9.1.5 Lithuania

The available studies, mostly, focus on food wastage from industries that generate the biggest amount, most of which comes from the dairy industry. The share of food waste is unknown.

Retailers are aware of their food waste and its causes. A possibility of voluntary donation is relatively low, because of willingness of companies to get also something back. This fact indicates on a necessity of the involvement of policy makers and development of incentives to stimulate this type of food waste treatment.

The Lithuanian Food Bank has launched campaign aimed at reduction of food waste, the only activity of this type, in the country. The organisation is also involved in related studies conducted at the national level.

Despite the available technical capacities for treatment of by-products, there are evidences of ineffective usage, as a result, only some parts are used for animal feed and biogas production.

Limited government participation 'enforces' some companies of the food industry to tackle the problem of their food waste treatment alone. On the other hand, such situations could be considered as an additional incentive for companies to reduce the amount of their food waste.

The share of renewable energy primary production from biomass and waste is one of the highest among the discussed countries. Therefore, it is most probably that food waste, together with other fractions of bio-waste, diverted from landfills, is used for this purpose.

9.1.6 Poland

The food waste situation differs from the situation in the rest of the countries, at large extent, and contradicts to the general conclusions about causes and methods of food waste reduction. The country is the second, after Germany, that generates the biggest amount of food wastage as well as of household food waste. Almost the third of total population are people at risk of poverty or social exclusion. Furthermore, the share of CAP expenditures on Food Programs in Poland in 2011 was more than 10 times bigger, than in any of the discussed countries.

The manufacturing sector generates the biggest amount of food wastage, mostly, because of obsolete facilities. With that, as in other countries, the share of food waste in that amount is unknown. At the same time, the country has a long experience in utilisation of by-products for animal feed and production of fertilisers.

The significant difference in the findings regarding the amount of food wastage generated by the manufacturing sector puts into question the accuracy of the statistical data. It might be caused by different classifications of food related waste types and their accounting.

One of the main methods of food waste reduction is donation. The main actor in the field, the Polish Federation of Food Banks, has an established cooperation with companies of the industry. The organisation initiates and supports different campaigns, projects and initiatives, aimed at raising public awareness on food waste and methods of its reduction. Food service companies demonstrate their interest in the problem and take steps to reduce food waste accumulated in their kitchen.

The analysis of the results on consumer knowledge about the food waste consequences refutes the prevailing opinion that the lack of such is one of the causes of household food waste.

Since the country is struggling to achieve the EU waste targets; most evidently that food waste together with other types of bio-waste is/will be diverted from landfills, and treated with available methods.

9.1.7 Sweden

The problem of food waste has become a research field several decades ago. In contrast to other countries, the specificity of the studies, as well as successfully implemented projects is narrow focused, i.e. single organisation, company or institution. It is hardly known if these practices are/will be introduced and implemented across the country.

The share of food waste, generated by food service institutions, schools and retailers, is much higher in comparison to the share of household food waste, while the latter is lower than that in Germany, and higher than the value stated in the WRAP studies.

The cooperation between representatives of different sectors including the government has a key role in all food waste related activities in the country. In addition to the implemented projects, there are also a variety of publications and guidelines regarding food waste management. Moreover, the reduction and treatment targets are set in the national management plan.

The main types of discarded food products, as well as the main causes of food waste are the same, as in other countries: fruits, vegetables, bakery and dairy products, and date labels, oversupply, improper handling of products and challenges to predict consumers demand respectively.

In contrast to other countries, the issue of food banks is not well developed. There are companies, particularly retailers that prevent/reduce their food waste with their own means, in some cases it also includes donation. Others prefer to send their food waste directly to treatment plants.

The treatment methods from the lower part of the waste hierarchy together with all required pre-treatment activities and infrastructure are developed very well. This situation makes no incentives for companies and households to consider other types of food waste treatment, especially, from the upper part of the food waste hierarchy. Moreover, the promotion of food wastage as one of the main sources for energy production and its associated positive effects makes reduction and prevention of food waste even less attractive.

9.2 Recommendations

In the last few years, the pool of recommendations regarding possible treatment methods of food waste has increased together with interest on to the problem. However, a peculiarity of each country, reflected in its economic and social situations, perception of food value in a society, consumer behaviour and etc., defines not only methods but also their necessary customization and adaptation.

Based on the analysis of available food waste related studies on the global level and in the discussed countries, existing treatment methods and best practices, the following recommendations could be made, in order to tackle the problem of food waste in the Baltic Region countries. There are a number of steps that needed to be implemented at the European level and be legally binding for all MS. It will not only increase the efficiency of the currently applied and future methods, but also allow better identification of the target area, e.g. particular stages in the food supply chain.

These steps are:

- Development of one single definition of the term 'food waste', including classification of food related waste types, generated along all stages of the food supply chain, and that could be referred as food waste.
- Increase of a number of undertaken studies on the national levels, in order to get detailed information, and accurate and as full as possible data about amounts of food waste generated at each stage of the food supply chain in each country.
- Setting of a reduction target for each country, considering its current food waste management situation, in the similar way, as it has been done for biodegradable waste.
- Development of guidelines for food waste data collection and reporting.
- Development of legally binding requirements for MS regarding the application of food waste treatment methods, particularly, from the upper part of the food waste hierarchy, to 'enforce'/motivate most of the countries to take concrete measures to tackle the problem.

The analysis of the food waste situation shows that despite existing differences, the implementation of the following measures is relevant for all discussed countries:

- In addition to awareness raising campaign, a launch of education campaigns about food date labels, food safety, correct shopping planning, usage of leftovers, appropriate storage, and dependence between, for example, a fruits/vegetables shape and quality of a product. Education campaigns that focus on understating of food value, an amount of resources required to produce/grow a product, a number of hungry people in the world, or at least in person's own country.

In regard with the retail industry:

- Limitation of an allowed number and type of marketing instruments that motivate consumers to buy excessive amounts of food.

- Distribution of practices aimed at reduction of food waste, such as 'buy one now, get one free later', and/or offering more products in bulk.
- Optimisation of ordering systems and avoidance of incorrect orders. Training schemes and workshops for staff members (Kranert et al. 2012), despite the recommendation was made for Germany, it is apparent to be relevant for all countries, as well.
- Generation of knowhow and the provision of advice on resource-efficient management and optimised material flow management (Kranert et al. 2012).

The following steps are recommended to take in order to improve quality of data:

- To oblige all types of companies that generate food waste to add a food waste section in their annual reports, where they will state not only a generated amount, but also methods of its treatment.
- In order to estimate and assess developments in the prevention of food waste and relevant action, constant collection of data and monitoring are essential (Kranert et al. 2012).
- To improve cooperation not only in a country, but also among countries, it is necessary to create a platform for communication, share of knowledge, experience and best practices.

Below, there is a list of recommendations that consider the current state of food waste problem in each country.

9.2.1 Belarus

- Preliminary studies and of an analysis of statistical data that come from organisation, which today are not available to the public.
- Implementation of methods aimed at diverting biodegradable waste from landfills. The process should be realised together with countrywide adaptation of separate collection of an organic fraction of MSW.
- Education programs and targeted information campaigns, aimed the general public, to raise awareness on biodegradable waste, food wastage and food waste generation. The current situation in the country makes it senseless and, even, to some extent, absurd to launch this type of activities only focused on the problem of food waste.
- Necessary involvement of the government, as of the main initiator and supporter of all types of activities aimed at food waste reduction.
- A need to find additional financial resources to change the food waste related situation.

The recommendation to increase tariffs for waste disposal is pointless, because of very high sensitivity of households to any increase of tariffs (e.g. electricity, water, waste). Moreover, the financial gap is covered by the waste tariffs for industries, so, in some way, this creates an incentive for industries to reduce their amounts of waste, because, supposedly, the waste tariffs for industries are relatively high.

9.2.2 Estonia

- Launch of public awareness campaigns and education initiatives/activities to fill the gap in information about date labels and food safety.
- Targeted information and educational campaigns aimed the manufacturing sector.
- Distribution of the information among companies about possible opportunities, including financial and legal aspects, to reduce food waste, ways of cooperation with charity organisations.
- Education activities regarding separate collection, environmental and financial consequences of food waste.
- Establishment of cooperation with companies of the food industry.
- Changes in legislation, in order to simplify and stimulate process of food waste reduction/prevention, for example, regulations related to food donation.
- Development of financial incentives for food waste reduction or prevention, e.g. changes of tax system—tax cuts, bonuses, an increase of waste tariffs.
- To include the issue of food waste into the national environmental strategy.
- Active government involvement and participation.

9.2.3 Germany

- Studies on effectiveness and efficiency of the already undertaken measures, as well as more detailed studies, focused on causes of household food waste.
- Studies to identify legal and logistical barriers to food waste reduction/prevention and development of possible approaches to overcoming these barriers (Kranert et al. 2012).
- Studies to close existing gaps in data and knowledge on the state of food waste in companies, with a focus on generated amounts and applied methods of treatment.
- Research on dependence between food donation to the German Food Banks and companies' (e.g. retailers, producers) motivation to reduce/prevent their food waste before donation.
- More education campaigns aimed at changing the perception of food value by households.
- Introduce separate collection for green and kitchen waste, it might help to visualise more prominently, amounts of thrown food.

- To introduce a relatively high tariff for kitchen waste, and possibly, to reconsider the method of its calculation, i.e. instead of per volume payments, use payments per kilogram.
- The active involvement of all relevant stakeholders in the food chain (agriculture, industry, trade, households, restaurant and catering sector, policy-makers, educational establishments, social institutions, etc.) (Kranert et al. 2012).

9.2.4 Latvia

- Initiate projects/campaigns aimed at reduction of food waste in the food industry. It might include information campaigns, training schemes for staff members, workshops, presentation and clarification of possible ways/methods of food waste reduction, as well as of related legal aspects and financial incentives.
- Information and education campaigns for the general public aimed at raising awareness about biodegradable waste, including the issue of food wastage and food waste, a necessity and importance of separate collection, environmental and other consequences.
- The active involvement and support of the government institutions.
- Establishment of stronger cooperation between charity organisations (e.g. Food Bank) and companies, involvement of a bigger number of companies.
- Monitoring of realisation of the binding regulations regarding companies' cooperation with organic waste service companies.
- Detailed studies on activities of this type of companies, particularly, collected amounts, methods of waste treatment and prices.

9.2.5 Lithuania

- Development of a unified waste classification system, at least between waste producers, collectors and waste treatment companies.
- Undertake studies, focused on food waste, generated by the meat, fish, dairy and bakery industries, in order to identify exact amounts of waste, its composition, and possible avoidable share, as well as causes.
- Undertake studies to identify reasons of non-efficient application of available technical capacities for processing of by-products.
- Undertake studies focused on household food waste, as well as, on food waste generated by the hospitality, retail and food service providing sectors. To identify quantities, causes and treatment methods.
- Launch information and education campaigns for the general public, about biodegradable waste, including the issues of separate collection, food wastage and food waste, environmental and other consequences.

- Launch information and education campaigns, as well as trainings and workshops for companies, to help them to treat generated food waste.
- Development of stronger collaboration between the food industry and companies, which provide waste management related services.
- Development of financial incentives to stimulate and support food donation.
- Extend the government involvement.

9.2.6 Poland

- Studies on the currently used waste classification system.
- Studies on food wastage generated by the manufacturing sector, including causes, treatment methods, as well as, a share of food waste.
- Studies regarding the state of the problem of food waste in other sectors (hospitality, retail, and etc.), with focus on generated amounts and methods of utilization.
- Studies to analyse the effectiveness of food donation to the Polish Food Banks, as of a method of food waste reduction/prevention.
- Development of regulations, financial incentives and etc. to motivate companies to modernise their technical potential.
- Development of additional legal and financial instruments aimed at stimulation of food waste reduction/prevention.
- Development of a platform for communication between different actors, e.g. NGOs, food producers, distributors, restaurants and food farms.
- Launch of projects, information and education campaigns for companies and organisations to fill the gap in knowledge about legal and financial aspects regarding food donation.
- Expand infrastructure for separate collection of bio-waste, to reduce the share that is landfilled. Enhance exchange of information/experience of municipalities already realising separate collection of bio-waste to other municipalities (BiPRO 2011).
- Promote home composting. Support the creation of a market for compost (BiPRO 2011).
- Introduce separate collection of kitchen waste and set a relatively high tariff rate for this type of waste.
- More active government participation in the food waste related activities.

9.2.7 Sweden

- Promotion and implementation of practices that were successfully realised in a single entity, across the country.
- Promotion and support of sharing experience in preventing food waste, but only measures from the upper part of the food waste hierarchy. Try to eliminate

situations when achievements in the food waste reduction process become the issue of competitive advantage of one company.
- Discourage retailers to send their food waste to treatment plants. Make this option least preferable/desirable, for example, by increasing costs of such activities.
- Take actions aimed at changing the consumer behaviour and demands (Stenmarck et al. 2011).
- Launch more national wide campaigns regarding household food waste. The campaigns should focus on methods of its prevention, as well as on environmental consequences, that in the case of Sweden, it relates to the problem of unsustainable use of resources.

To reduce food waste in the hospitality sector, experts recommend: (Marthinsen et al. 2012)

- Develop routines for right portions.
- Conduct internal trainings on costs (cost of food/total costs).
- Launch general awareness campaigns on food waste prevention.
- Develop better menu planning.

In conclusion, the undertaken research has shown that the problem of food waste is not thc problem of only developed or only undeveloped countries. From policy-makers to industrialists and NGOs, many actors are involved in the issue of food waste management, and the solutions required are cross-sectoral. An economic situation in a country and/or government indifference is not a reason to ignore the problem. Apart from the significant financial and environmental losses, it is the social and moral responsibility of countries to tackle the problem in the world, where according to the 'State of Food Insecurity in the World' report (FAO et al. 2012), childhood malnutrition is a cause of death for more than 2.5 million children every year.

References

BiPRO (2011). *Roadmap for Poland (PL)*, Retrieved from http://ec.europa.eu/environment/waste/framework/pdf/PL_Roadmap_FINAL.pdf.

FAO, WFP & IFAD (2012). *The State of Food Insecurity in the World 2012. Economic growth is necessary but not sufficient to accelerate reduction of hunger and malnutrition*, Rome. Retrieved from http://www.fao.org/docrep/016/i3027e/i3027e.pdf.

Kranert, M. et al. (2012). *Determination of discarded food and proposals for a minimization of food wastage in Germany. Abridged Version*, Stuttgart. Retrieved from http://www.bmelv.de/SharedDocs/Downloads/EN/Food/Studie_Lebensmittelabfaelle_Kurzfassung.pdf;jsessionid=D16102240D2BF5C765C01411351A57ED.2_cid367?__blob=publicationFile.

Marthinsen, J. et al. (2012). *Prevention of food waste in restaurants, hotels, canteens and catering*, Copenhagen. Retrieved from http://www.norden.org/en/publications/publikationer/2012-537.

Stenmarck, Å. et al. (2011). *Initiatives on prevention of food waste in the retail and wholesale trades*, Copenhagen. Retrieved from http://www.ivl.se/download/18.7df4c4e812d2da6a416800089028/B1988.pdf.

Appendix A

Questionnaire in English

W. Leal Filho and M. Kovaleva, *Food Waste and Sustainable Food Waste Management in the Baltic Sea Region*, Environmental Science and Engineering, DOI 10.1007/978-3-319-10906-0

Questionnaire

This questionnaire is a part of a book which centers on the problem of food waste in countries in the Baltic region.

For the purpose of the book the term 'foodwaste'is defined asavoidable waste atevery stage ofthe food system, where discarded food has stilla value and fits to consumption.

Please indicate your country: ______________

Please answer the following questions:

1. *In your country:* Is there a separate collection of bio-waste and food waste or food waste is considered as a part of bio-waste and disposed together?

   ○ Yes

   ○ No

2. *In your country:* How households and companies pay for bio-waste or for food waste in the case of its separate collection:

   ☐ By volume (cubic meter/liters)

   ☐ By weight

   ☐ Fixed monthly payment

   ☐ Other (please specify):

3. *In your country:* Is the price for disposal of bio-waste or food waste higher or lower than the price for disposalo fo ther waste types, in the case if there is separate collection of waste?

   ○ Higher

   please specify, for how much

   ○ Lower

   please specify, for how much

4. *In your country:* What type of promotional tools (promotions) are frequently used in the grocery sector?

☐ Multi-buy e.g. 3 for 2, 5 for 4 - Any multiple number of items that are sold with another free

☐ BOGOF (Buy one get one free)

☐ Extra free- an increased size of an item whilst maintaining the current cost

☐ Y for €x - a number ofitems for a set amounti.e. 2 for €3, 3 for €6

☐ TPR - Temporary Price Reduction e.g.initial price reduced by 50%

☐ Other (please specify):

5. *In your country:* How easy is to buy products in bulk, for example flour, noodles, sugar, rice and etc.?

☐ In every super market

☐ Only in particular stores

☐ Impossible to buy

☐ Other (please specify):

6. *In your country:* What is the share of food products produced locally, including agricultural produce?

○ less than15%

○ between 15% and 30%

○ between 30% and 50%

○ between 50% and 80%

○ more than 80%

7. What are the main causes of food waste in your country? Please name at least 3 main causes

8. Currently, what are the main methods of bio- waste and food waste treatment in your country:

☐ Incineration

☐ Recycling

☐ Composting

☐ Landfill

☐ Other (please specify):

9. *In your country:* What is the extent of usage of Anaerobic Digestion (AD) plants?

○ Restricted

○ Wide

○ Very wide

10. *In your country:* are there dedicated AD food waste treatment plants?

   ◯ Yes

   ◯ No

   If yes, how many?

11. In your opinion, in your country, at what stage/stages of the food system there is a biggest share of food waste generation?

12. *In your country:* At what level the issue of food waste is considered as a problem:

   ☐ National (ministries)

   ☐ Local – single governmental workers

   ☐ Only NGOs and activists

   ☐ Other (please specify):

13. Are there measures taken by the government in your country to solve the problem of food waste?

14. *In your country:* is the problem of food waste reflected in the National Waste Prevention Program?

   ◯ Yes

   ◯ No

   If yes, at what extent/in which context?

15. If there are any widely spread awareness-raising food waste reduction campaigns conducted in your country (for example, like Love Food Hate Waste in the UK)?

   ◯ Yes

   ◯ No

   If yes, please specify

16. What type of methods/initiatives/good practices of food waste reduction exist in your country? Please specify.

17. *In your country:* if there is mandatory or voluntary reporting o f food waste by companies (e.g. producers, retailers, restaurants and etc.)?

○ Yes

○ No

If yes, please specify

18. In your opinion, which of the following methods and/or instruments of food waste reduction would be the most effective in your country:

☐ Awareness campaigns among consumers/industry

☐ Educational programs

☐ Extended producers/companies responsibility

☐ More informative labelling (incl.guidelines for storage, additional info about food waste)

☐ Promotions (e.g. buy one now,one later)

☐ Positive and/negative financial stimuli (fines, taxes, subsidies)

☐ Industry voluntary agreements to reduce amount of food waste

☐ Other (please specify):

19. In your opinion, what are the main barriers/obstacles for effective implementation of activities aimed at reduction of food waste generation?

20. In your opinion, is there a need in a separate legislative documents/legislation for food waste?

○ Yes

○ No

If yes, please specify

Thank you very much for your time and participation.

Appendix B

Questionnaire in Russian

W. Leal Filho and M. Kovaleva, *Food Waste and Sustainable Food Waste Management in the Baltic Sea Region*, Environmental Science and Engineering, DOI 10.1007/978-3-319-10906-0

Опросник

Данный опросник является частью книги, посвященной проблеме необоснованных пищевых отходов в странах Балтийского региона, в том числе и в Беларуси.

В рамках данной работы термин «**необоснованные пищевые отходы**» определен как необоснованные отходы продуктов питания возникающие на каждом этапе продовольственной цепочки, которые все еще могли бы быть пригодны для потребления.

Пожалуйста ответьте на следующие вопросы:

1. *Осуществляется ли раздельный сбор биологических и/или пищевых отходов в вашей стране*?

   ○ Да

   ○ Нет

2. *В случае раздельного сбора биологических и/или пищевых отходов, в основу расчета платы на вывоз данного типа отходов ложится*:

   ☐ Объем (кубический метр/литр)

   ☐ Вес

   ☐ Фиксированная ежемесячная сумма)

   ☐ Другое (пожалуйста поясните):

3. *В вашей стране, в случае раздельного сбора отходов, размер оплаты за вывоз биологических и/или пищевых отходов выше или ниже размера оплаты за вывоз других типов отходов*?

   ○ Выше

   Пожалуйста укажите насколько

   ○ Ниже

   Пожалуйста укажите насколько

4. *Какие рекламные акции, направленные на увеличение сбыта продуктов питания, наиболее часто используются в продовольственном секторе в вашей стране?*

   ☐ Бонусный набор: два по цене одного; три по цене двух и т.д.

   ☐ Увеличенная упаковка по прежней цене (напр. +10% бесплатно)

   ☐ N кол-во шт. за фиксированную сумму, напр. 2 за 300 руб., 3 за 600 руб.

☐ Временные скидки

☐ Другие (пожалуйста укажите):

5. *В вашей стране, насколько легко можно купить продукты питания на разве*
например, сахар, рис, муку, макаронные изделия и т.д.?

☐ В любом магазине/супермаркете

☐ Только в определенных магазинах

☐ Невозможно купить

☐ Другое (пожалуйста укажите):

6. *В вашей стране, какая часть продуктов питания, включа*
сельскохозяйственную продукцию, является местного производства?

○ меньше 15%

○ между 15% и 30%

○ между 30% и 50%

○ между 50% и 80%

○ больше 80%

7. *Пожалуйста назовите по меньшей мере 3 основные причины образовани*
необоснованных пищевых отходов в вашей стране?

8. *Какие основные методы обработки/утилизации биологических и пищевы*
отходов существуют в вашей стране на сегодняшний день:

☐ Сжигание

☐ Рециклинг

☐ Компостирование

☐ Захоронение

☐ Другие (пожалуйста укажите):

9. *Насколько широко в вашей стране распространено использование установок дл*
анаэробного сбраживания?

○ Используется только необольшое число установок

○ Широко используются

○ Повсеместно

10. *Используются ли в вашей стране установки для анаэробного сбраживани только пищевых отходов (любого типа)?*

◯ Да

◯ Нет

Если да, то сколько?

11. *По вашему мнению, в вашей стране, на каких этапах продовольственно цепочки образуется наибольшее количество пищевых отходов, в том числе необоснованных?*

12. *В вашей стране, на каком уровне управления вопрос необоснованных пищевы отходов считается проблемой, требующей незамедлительного решения?*

☐ Национальный (на уровне министерств)

☐ Местный - сотрудники муниципальных госучереждений

☐ Только НПО и активисты

☐ Другой (пожалуйста укажите):

13. *В вашей стране, на сегодняшний день, существуют ли государственные мер по решению проблемы необоснованных пищевых отходов?*

14. *Отражена ли проблема необоснованных пищевых отходов в государственно программе, касающейся сокращению/предотвращению отходов в вашей стране*

◯ Да

◯ Нет

Если да, то в каком контексте, в какой степени?

15. *Проводятся ли в вашей стране информационно-образовательные кампани направленные на сокращение количества необоснованных пищевых отходо например, такие как «Love Food Hate Waste» в Великобритании?*

◯ Да

◯ Нет

Если да, пожалуйста приведите примеры

16. *Какие методы/инициативы по сокращению необоснованных пищевых отходс существуют в вашей стране? Пожалуйста уточните*

17. *В вашей стране, существует ли для компаний (производителей, торговых сетей, ресторанов и т.д.) обязательная или добровольная форма отчетности, касающаяся количества и типа образованных пищевых отходов?*

○ Да

○ Нет

Если да, пожалуйста уточните

18. *По вашему, мнению какие из перечисленных ниже методов/инструментов сокращения количества необоснованных пищевых отходов были бы наиболее эффективными в вашей стране:*

☐ Информационные кампании для потребителей и представителей индустрии

☐ Образовательные программы

☐ Расширенная ответственность производителя/продовца

☐ Более полная информация на этикетках, в том числе, информация о пищевых отходах

☐ Акции напр. один купи сейчас, другой получи позже

☐ Положительный и/или отрицательные финансовые стимулы (напр. налоги, штрафы, субсидии)

☐ Добровольные соглашения между представителями индустрии по уменьшению количества необоснованных пищевых отходов

☐ Другие (пожалуйста укажите):

19. *По вашему мнению, какие основные барьеры/трудности существуют в вашей стране для эффективного внедрения мер направленных на сокращение количества образующихся необоснованных пищевых отходов?*

20. *По вашему мнению необходимо ли отдельное законодательство/ нормативные акты/ инструкции и т. д., касающиеся пищевых отходов, в том числе и необоснованных?*

○ Да

○ Нет

Если да, пожалуйста укажите

Большое спасибо за ваше время и участие.

Glossary

Anaerobic digestion is a process where microorganisms break down organic materials, such as food scraps, manure, and sewage sludge, in the absence of oxygen

Animal by-products are entire bodies or parts of animals or products of animal origin not intended for human consumption, including ova, embryos and semen

Biofertiliser is a natural fertilizer that helps to provide and keep in the soil all the nutrients and micro-organisms required for the benefits of the plants

Biofuels are gas or liquid fuel made from plant material (biomass). Includes wood, wood waste, wood liquors, peat, railroad ties, wood sludge, spent sulphite liquors, agricultural waste, straw, tires, fish oils, tall oil, sludge waste, waste alcohol, municipal solid waste, landfill gases, other waste, and ethanol blended into motor gasoline

Biogas is mixture of gases produced by anaerobic digestion

Biological treatment includes composting and anaerobic digestion, may be classified as recycling, when compost (or digestate) is used on land or for the production of growing media

Biomass is materials that are biological in origin, including organic material (both living and dead) from above and below ground, for example, trees, crops, grasses, and animal waste

Bio-waste means biodegradable garden and park waste, food and kitchen waste from households, restaurants, caterers and retail premises and comparable waste from food processing plants

BOGOF (Buy one get one free) is a form of multi-buy that is split out from other multi-buys

By-product is a substance or object, resulting from a production process, the primary aim of which is not the production of that item, meeting the following conditions: (a) further use of the substance or object is certain; (b) the substance or object can be used directly without any further processing other than normal

W. Leal Filho and M. Kovaleva, *Food Waste and Sustainable Food Waste Management in the Baltic Sea Region*, Environmental Science and Engineering, DOI 10.1007/978-3-319-10906-0

industrial practice; (c) the substance or object is produced as an integral part of a production process; and (d) further use is lawful.

Co-digestion is a process, whereby one or more waste types are digested in a mixture, in order to enhance digester efficiency and increase biogas yield

Collection is the gathering of waste, including the preliminary sorting and preliminary storage of waste for the purposes of transport to a waste treatment facility

Compost is the odourless, stable and humus-like material rich in organic matter, as well as proteins and carbohydrates, issued from the composting process of biodegradable wastes

Composting is the aerobic decomposition of biodegradable wastes under controlled conditions and their reconstitution into humus by the action of micro- and macro-organisms, involving the bonding of nitrogen onto carbon molecules, fixing proteins and carbohydrates in forms readily available to plants

Culling is the process of the removal of products based on quality or appearance criteria, including specifications for size, colour, weight, blemish level and etc.

Denaturation is a process, when substances are added to food wastage, in order to change its properties and prevent its future use for feeding purposes

Digestate is a nutrient-rich material left following anaerobic digestion

Digester is the tank in which anaerobic digestion takes place

Disposal is any operation, which is not recovery, even, where the operation has as a secondary consequence the reclamation of substances or energy

Ecosystem is any natural unit or entity including living and non-living parts that interact to produce a stable system through cyclic exchange of materials

Emissions are the release of a substance (usually a gas when referring to the subject of climate change) into the atmosphere

Extraction rates is the concept that relates to processed products only and indicates, in per cent terms, the amount of the processed product concerned obtained from the processing of the parent/originating product, in most cases a primary product. For example, flour from wheat, oil from soybeans, shelled from unshelled almonds, cheese from milk, etc

Fertiliser is a substance added to soil to make it more fertile

Food losses are wholesome edible material intended for human consumption that is instead, lost as an unintended result of agricultural processes, lack of technology or technical limitations in storage, packaging, and/or marketing, poor infrastructure and logistics, insufficient skills, knowledge and management capacity of supply chain actors or consumed by pests

Food recovery is a collection, or recovery, of wholesome food from farmers' fields, retail stores, or foodservice establishments for distribution to the poor and hungry

Food redistribution is voluntarily giving away of food that otherwise would be lost or wasted to recipients e.g. charitable organisation, which then redistribute the food to those who need it

Food residuals are unavoidable inedible and partly avoidable portions such as skins, bones, stalks, shells and leaves

Food system is all processes involved in providing food and food-related items to a population, including growing, harvesting, processing, packaging, transporting, marketing, consumption and disposal

Food wastage are all types of food/food products that were produced (e.g. grew, manufactured, cooked) for human consumption, however, thrown away. It includes such types of discarded food as food losses, food residuals, by-products, including animal by-products and food waste

Food waste is the category of avoidable waste, when discarded food has still value and very often fits to consumption. Food waste is food that is spilled, spoiled, bruised or wilted. It may include whole or unopened packs or individual items of foods which are not eaten at all. Food waste arises at any point in the food supply chain as a result of inappropriate behaviour of food chain actors e.g. producers, retailers, the food service sector, consumers as well as of a lack of existing technologies

Fresh food includes fresh fruit, vegetables, salad items, herbs, bread, milk and dairy products, meat and seafood

Frozen food includes frozen vegetables and fruit, chips, ready-made meals and frozen desserts

Gasification is any chemical or heat process used to convert a substance to a gas. The process takes place at high temperature. The gasification product is a mixture of combustible gases and tar compounds, together with particles and water vapour

Greenhouse gas (GHG) is any gas that absorbs infrared radiation in the atmosphere. Greenhouse gases include, carbon dioxide, methane, nitrous oxide, ozone, chlorofluorocarbons, hydrochlorofluorocarbons, hydrofluorocarbons, perfluorocarbons, sulfur hexafluoride

Hazardous waste is any waste or combination of wastes with the potential to damage human health, living organisms or the environment. Hazardous wastes usually require special handling and disposal procedures which are regulated by national and international laws

Home composting scheme is a scheme, when biodegradable waste generated by householders is used to produce compost for use by the individual

Household is one person living separately or a group of people who live in a common main dwelling and share joint financial and/or food resources

Incineration is the controlled burning of solid, liquid, or gaseous combustible wastes to produce gases and solid residues containing little or no combustible material in order to reduce the bulk of the original waste materials

Industrial waste is waste generated in the process of economic activity of legal persons and individual entrepreneurs (manufacture of goods, electricity generation, performing of work, provision of services), by- and associated products of mineral extraction and processing

Infrastructure is the technical structures that support a society, such as roads, water supply, sewerage, power grids, telecommunications

In-vessel composting is a process, when biodegradable material is composted inside a drum, silo, container or other structure

Landfill gas is gas generated in landfill sites by the anaerobic decomposition of domestic refuse (municipal solid waste). It consists of a mixture of gases and is colourless with an offensive odour due to the traces of organosulphur compounds

Landfill is a waste disposal site for the deposit of the waste onto or into land (i.e. underground)

Leftovers are any uneaten food portions or ingredients remaining from a previous meal that can be eaten at a later date including take away meals, home cooked dinners or individual cooked ingredients like pasta

Material life cycle is all the stages involved in the manufacturing, distribution and retail, use and re-use and maintenance, recycling and waste management of materials

Maximum residue levels (MRLs) are the upper legal levels of a concentration for pesticide residues in or on food or feed based on good agricultural practices and to ensure the lowest possible consumer exposure

Mechanical-biological treatment (MBT) are techniques, which combine biological treatment with mechanical treatment (sorting)

Methane (CH_4) is a hydrocarbon that is a greenhouse gas with a global warming potential most recently estimated at 25 times that of carbon dioxide (CO_2). Methane is produced through anaerobic (without oxygen) decomposition of waste in landfills, animal digestion, decomposition of animal wastes, production and distribution of natural gas and petroleum, coal production, and incomplete fossil fuel combustion

Multi-buy is any multiple number of items that are sold with another free (e.g. 3 for 2, 5 for 4)

Municipal waste is waste from households, as well as other waste which, because of its nature or composition, is similar to waste from households

Packaged and long life food is sweet and savoury biscuits, chips, rice, cereal, flour, coffee and tinned food

Perishable food is meats, dairy products, produce, and bakery item s that are donated from grocery stores, produce distributors, food distributors, etc

Prepared foods are foods of all descriptions that have been prepared but were never served. This includes cooked items, such as meats, entrees, vegetables, starches, deli trays, and vegetable trays

Prevention is measures taken before a substance, material or product has become waste

Recovery is any operation the principal result of which is waste serving a useful purpose by replacing other materials which would otherwise have been used to fulfil

Recycling is any recovery operation by which waste materials are reprocessed into products, materials or substances whether for the original or other purposes. It includes the reprocessing of organic material but does not include energy recovery and the reprocessing into materials that are to be used as fuels or for backfilling operations

Renewable energy is energy resources that are naturally replenishing such as biomass, hydro, geothermal, solar, wind, ocean thermal, wave action, and tidal action

Re-use is any operation by which products or components that are not waste are used again for the same purpose for which they were conceived

Separate collection is the collection where a waste stream is kept separately by type and nature so as to facilitate a specific treatment

Supply chain is a system of organisations, people, technology, activities, information and resources that begins with the sourcing of raw material and extends through the delivery of end items to the final customer

Take-back is a system whereby, some retailers include clauses in supply contracts that entitle them to return stock to their suppliers once it has reached a specified amount of residual shelf-life remaining e.g. 75 %

TPR Temporary Price Reduction e.g. initial price reduced by 20 %

Treatment is recovery or disposal operations, including preparation prior to recovery or disposal

Waste is any substance or object which the holder discards or intends, or is required to dispose of pursuant to the provisions of national law in force

Waste management is the collection, transport, recovery and disposal of waste, including the supervision of such operations and the after-care of disposal sites, and including actions taken as a dealer or broker

Windrow composting is a process that usually relies on natural processes for air supply to the waste, although it may be artificially aerated. Windrows are turned to increase the porosity of the pile, and increase the homogeneity of the waste

Windrows regularly turned elongated piles of waste in the process of being composted